Vanessa Antunes de Souza
Cenira M. Lupinacci
Darlene A. F.Oliveira

Multifunctionality of the rural landscape

AF671233

Vanessa Antunes de Souza
Cenira M. Lupinacci
Darlene A. F.Oliveira

Multifunctionality of the rural landscape

Study to identify and characterise the landscape and the multifunctional potential of the Itapé-Rio Claro (SP) rural district

ScienciaScripts

Imprint
Any brand names and product names mentioned in this book are subject to trademark, brand or patent protection and are trademarks or registered trademarks of their respective holders. The use of brand names, product names, common names, trade names, product descriptions etc. even without a particular marking in this work is in no way to be construed to mean that such names may be regarded as unrestricted in respect of trademark and brand protection legislation and could thus be used by anyone.

Cover image: www.ingimage.com

This book is a translation from the original published under ISBN 978-3-330-76056-1.

Publisher:
Sciencia Scripts
is a trademark of
Dodo Books Indian Ocean Ltd. and OmniScriptum S.R.L publishing group

120 High Road, East Finchley, London, N2 9ED, United Kingdom
Str. Armeneasca 28/1, office 1, Chisinau MD-2012, Republic of Moldova, Europe
Printed at: see last page
ISBN: 978-620-8-33924-1

Copyright © Vanessa Antunes de Souza, Cenira M. Lupinacci, Darlene A. F.Oliveira
Copyright © 2024 Dodo Books Indian Ocean Ltd. and OmniScriptum S.R.L publishing group

ACKNOWLEDGEMENTS

Dr Darlene Aparecida de Oliveira, for her guidance and support, being a very important person for my training.

I would like to thank Dr Cenira Maria Lupinacci for her geomorphological teachings, guidance, opportunity and professional example.

To my siblings Ivone Antunes, Tatiane Antunes and João Paulo Antunes for their great partnership and friendship, who have always supported me unconditionally, helping, respecting and motivating me in difficult times, without whom my education would not have been possible.

I would like to thank Tiago Felipe Cerri, for his love and companionship, for having walked with me during this period, helping, supporting, motivating and respecting.

To my friends and colleagues Adriana Moraes, Valéria Bernadelli, Talita Jordão and Sheilla Ribeiro for the moments of friendship, help, worries, laughter and construction. People who have been and will continue to be very important to me, as examples of competence, wit, sincerity, criticality and friendship. Thanks girls.

I would like to thank my friend and colleague Luiz Henrique Pereira for his great help and willingness, his friendship and his professional example to follow.

To the members of the Agrarian Studies Centre for their help, friendship, partnership in field work, events, activities and discussions that we carried out together. Thanks guys!

To my colleagues, teachers and staff at the Geography Department of UNESP Rio Claro, who have always been there for me.

To my colleague and friend Gilberto Donizete Henrique, a draughtsman at the Geography Department of UNESP in Rio Claro, for his friendship, willingness, help and friendliness.

To my colleagues at the Geomorphology Laboratory, the professors and staff at the Department of Territorial Planning and Geoprocessing at UNESP Rio Claro, for their help with the mapping work and for sharing with you such a great place to work and learn.

To CNPq for awarding me a scholarship. Finally, I would like to thank everyone who helped in any way with the research.

SUMMARY

The general aim of this study was to identify and characterise the landscapes and multifunctional potential of small rural properties in the Itapé rural neighbourhood, located in the catchment area of the Sapezeiro stream in the northwest of the municipality of Rio Claro. To this end, our theoretical approach was based on the four analytical elements highlighted by Diniz (1984, p.30 apud FERREIRA, p.50): rural landscape/multifunctionality/external elements of agriculture/endogenous rural development. In order to characterise the physical attributes of the landscape in the Itapé neighbourhood in Rio Claro - SP, a survey was carried out of topographical charts, maps and aerial images, as well as rainfall data that enabled the natural conditions of the area to be identified. Through direct research with field observation and the application of questionnaires, we sought to collect socio-economic data as well as the tourist and landscape resources that exist in the area that enable the development of agricultural and/or non-agricultural activities as alternatives for employment and income. The aim was to understand the limitations and difficulties in terms of socio-economic aspects, infrastructure and the physical attributes of the landscape that imply limited development for local families. In view of the information gathered, it was possible to assess the diversity and specificities of the physical and socio-economic elements of the area studied, identifying the presence of pluriactivity and multifunctionality of the rural space in the area surveyed.

Keywords: Multifunctionality; Rural landscape; Family farming; Rural development.

SUMMARY

CHAPTER 1

INTRODUCTION

With the expansion of agricultural activities without taking into account the potential and limitations of the land, rural landscapes have undergone drastic transformations in recent years, and such activities are a potential factor in environmental degradation. The aim of this study was to identify and characterise the landscapes and multifunctional potential of the rural district of Itapé, located in the north-west of the municipality of Rio Claro - SP, covering an area of approximately 23 km^2, with the geographical coordinates 22° 18'S and 22° 21'S and 47°39'We47° 45'W as references.

The study focused on the neighbourhood because of its spatial specificities, seeking to understand the limitations and difficulties in terms of both the physical and socio-economic aspects that imply the limited development of local families. The municipality of Rio Claro is characterised as urban-industrial and agricultural production is geared towards monoculture, which hinders the development of the smallholdings that predominate in the area studied.

In parallel with the absence of significant economic development, it is possible to look for new ways of using and occupying the area. Certain areas may lose their ability to compete in terms of production, but they can develop other functions such as support, recreation, nature conservation, environmental quality, aesthetics (especially for tourists, who will use the landscape for consumption) and preservation of cultural and architectural heritage. These functions will be more highly valued, since not all rural spaces will be productive spaces, but they can develop other functions linked to the post-productive demand for rural space.

We are currently seeing a growing demand for rural areas, both as a residential alternative (weekends and holidays) and as a tourist alternative, associated with ideologies that promote escaping from the city to the countryside (the latter understood as a refuge from the troubled environment that surrounds the city). Rural areas and their landscapes are, above all, reminiscent of tranquillity, roots, "authenticity", due to their peculiar characteristics, the natural environment and freedom, harmony and quality of life.

Working with the countryside as a space for living, working and leisure, and not just for agricultural production, opens up new possibilities for families to stay, guarantees food security, environmental protection, the preservation of tangible and intangible cultural heritage and, consequently, its development.

Thus, when assessing the potential for rural multifunctionality, one considers not only ways of maintaining family farmers, but also local development, which should allow for social and cultural reproduction and environmental maintenance, among other things. According to the OECD (2001, apud PINTO-CORREA 2008, 67-71), multifunctionality is the reconciliation of comforts and amenities "since production functions, commodities, are now joined by functions that are not directly productive, *non-commodities"*. In this case, our study area can reconcile *commodities* with socio-cultural, economic and environmental functions.

CHAPTER 2

METHODOLOGICAL PROCEDURES

As already mentioned, the aim of this work was to analyse the landscape and production potential of small rural properties in the Itapé rural neighbourhood, and in order to achieve these objectives the following characterisations were carried out:

- of the rural area of the municipality of Rio Claro with a view to identifying the dominant land use systems and the non-productive functions that are most important in the current situation;
- of farms, taking into account: size, main activity, level of income and investment, existing equipment on the property that allows for diversification of activities;
- of family producers, seeking to identify the group's economic, cultural, entrepreneurial and educational potential;
- and recognition of the rural landscape and the physical attributes of the Sapezeiro Stream Basin, where the Itapé neighbourhood is located.

2.1 Materials used

The following items were used to carry out the activities:

2.1.1 Mapping documentation survey

L Survey of Topographical Charts on a scale of 1:10,000 (Secretariat of Economy and Planning-Cartographic Plan of the State of São Paulo, 1979), articulations of the sheets:

- Jardim Ubá SF-23-Y-A-I-3-NE-D
- Itapé neighbourhood SF-23-Y-A-I-3-SE-B

- Bizigueli Hill SF-23-Y-A-I-3-SE-B
- Monjolo Grande stream SF-23-Y-A-I-4-NO-E

- Boa Vista neighbourhood SF-23-Y-A-I-4-NO-D
- Camacuã Forest Garden SF-23-Y-A-I-4-NO-F

M Thematic maps: Geological at a scale of 1:50,000 (IG,1986) and Pedological at a scale of 1:100,000 (OLIVEIRA E PRADO, 1981)

A Acquisition of orthorectified aerial photographs on a scale of 1:25,000 (EMPLASA-Projeto Mapeia São Paulo, 2011).

2.1.2 Geoprocessing system and other computer software

S Geographic Information System software ArcMap (ArcGis version 10.2);

S Database management systematisation, from Microsoft Office Access (2007);

E Microsoft Office Excel (2010) spreadsheet editor, making tables and graphs.

2.2 Methods

The following methods were used to achieve the objectives of this research:

A *A* theoretical and conceptual deepening of the landscape, smallholdings and multifunctionality;

D Development of simple observation and information gathering, recording in field notebooks and iconography in order to characterise the farming system and the rural landscape;

R Carrying out fieldwork and preparing forms to be applied to rural producers;

S Systematising information and organising it in a database;

T Transcription of the interviews;

T Data tabulation;

Analysing the results of the field research with the production of textual, image and digital material;

In view of the objectives defined and considered fundamental to this research, we will present the procedures carried out.

Initially, information was gathered by researching data available on the web and also in bibliographic material, in order to define a general picture of the agribusiness situation in Rio Claro. This was followed by an on-site reconnaissance of the research area to collect information and make notes in a field notebook of aspects that were deemed important for a better characterisation.

The forms were drawn up with the theoretical support of four analytical elements: rural landscape/multifunctionality/external elements of agriculture/endogenous rural development according to Diniz (1984, p.30 apud FERREIRA, p.50). Information was collected from 33 properties of up to 4 fiscal modules (approximately 50 hectares, INCRA, 2013[1]). These properties are considered to be small or family rural holdings, as per Article 3 of Law No. 11.326 of 24 July 2006.

During the visits, closed, empty properties were observed, not to mention the opposition of some residents to answering the forms. The fieldwork was also complemented by six interviews organised around open-ended, non-directive questions specifically formulated for people who have lived in the neighbourhood for longer (>50 years) and heirs, who are familiar with local changes and traditions.

2.3 Technical procedures

The information collected in the field was systematised, organised and entered into a database using Microsoft Office Access (2007) software, allowing for the filtering and selection of information deemed necessary through queries, and Microsoft Excel software for drawing up tables and graphs. The properties visited were also located using points plotted on a Google Maps satellite image. In this way, it was possible to obtain the geographical coordinates of these properties in degrees, minutes and seconds (DD.MM.SS), which were entered into the Excel software and converted into decimal geographical coordinates (DD.dddd), making it possible to enter them into the GIS (ArcGis 10.2) and spatialise the coordinates in point geometry (Figure

[1] According to the National Institute for Colonisation and Agrarian Reform, a Fiscal Module (ha) in Rio Claro - São Paulo corresponds to 14 hectares (INCRA, 2013).

1). These were the control points that helped us georeference the maps.

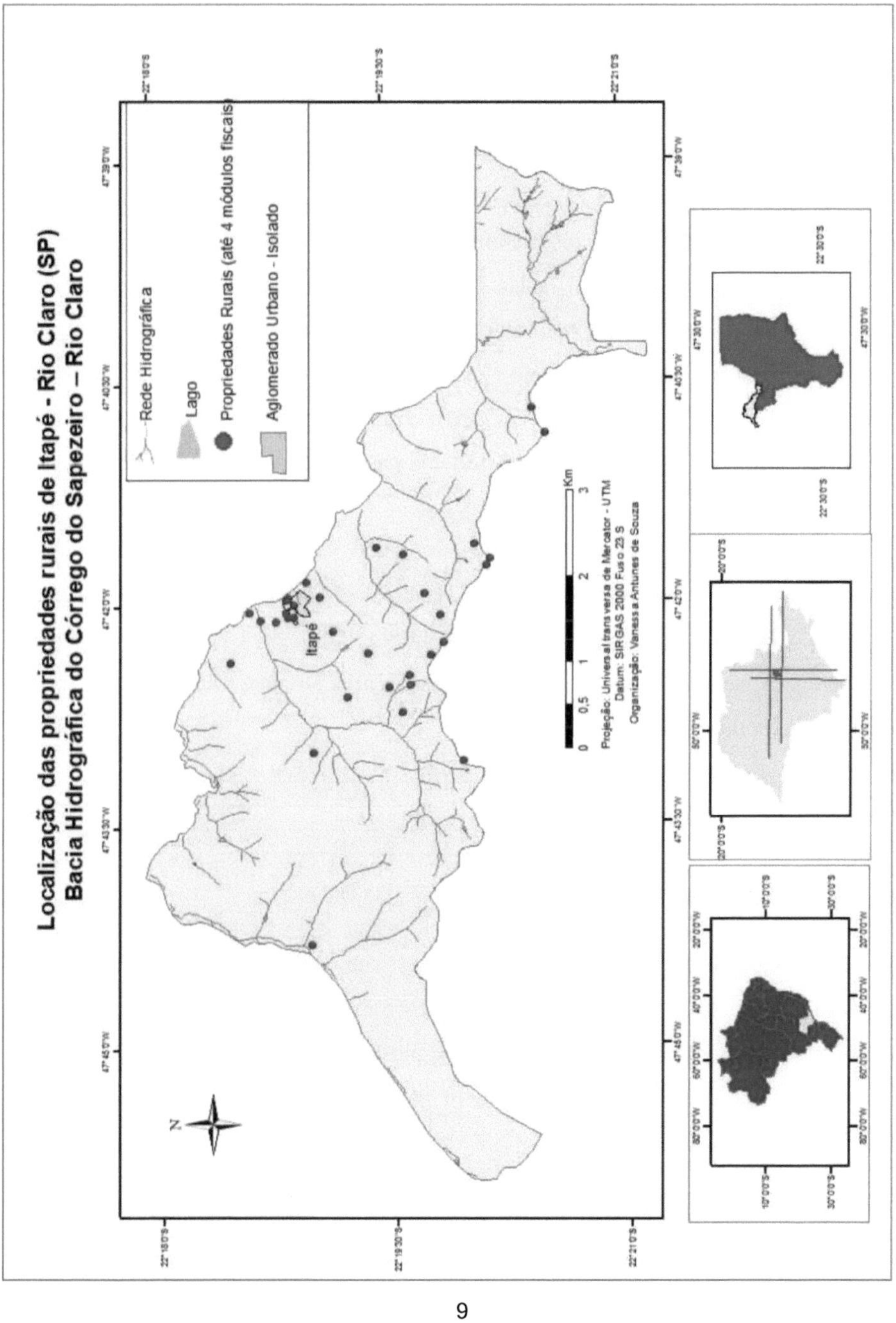

Figure 1 - Location of the rural properties visited Source: Organisation, V. A. Souza, 2016.

For the cartographic documentation of the Sapezeiro stream catchment area, where the Itapé neighbourhood is located, the scale adopted was 1:10,000 and 1:50,000 due to the fact that they allow for a greater level of detail of the study area.

In order to build the cartographic base, we needed topographic maps digitised at 300 dpi resolution in tiff format, which were provided by the Geography Department and DEPLAN. These were inserted into the GIS (ArcGis version 10.2) to be georeferenced, using the UTM (Universal Transverse Mercator) plane coordinate system and the SIRGAS 2000 Datum .[2]

For georeferencing, the Cartographic Accuracy Standard (PEC) was used, determined by Decree No. 89.817 of 20 June 1984, which sets out Regulatory Instructions for National Cartographic Technical Standards. Article 9 of the Decree states that maps must be classified according to their accuracy as Class A, B or C. According to Article 7, maps on a scale greater than 1:25,000 will have their articulation, format and projection system regulated by a separate standard.

Therefore, in order to classify the map accurately, you need to know what quality is required when determining reference points. This quality depends on the scale and class expected for the map. According to Merchant (1982 apud FERNANDES, 2009, p. 39), reference points should be determined using procedures in which the error is no more than 1/3 of the standard error expected for the class of the map (Figure 2).

Table 3: Planimetric accuracy standard and standard error for classes A, B and C

	Class A	C class B	Class C
Planimetric CSP (1)	**0.5 mm(1)**	**0.8 mm(1)**	**1.0 mm(1)**
Standard error(1)	**0.3 mm (1)**	**0.5 mm(1)**	**0.6 mm(1)**

Figure 2 - PEC
Source: BRASIL (1984) apud FERNANDES, 2009, p. 39. (1) Chart scale.

In view of this, the decision was made to work with the Sapezeiro Hydrographic Basin, whose area was delimited using material supplied by the Rio Claro City Council in *Shapefile (polygon)* format from the "Environmental Diagnosis and Development of Systems for Implementing Projects to Recover the Quality of Water Bodies" (ECosBio, 2013), which was

[2] SIRGAS 2000 (Geocentric Reference System for the Americas). IBGE Presidential Resolution No. 1/2005 establishes the Geocentric Reference System for the Americas (SIRGAS), in its 2000 realisation (SIRGAS2000), as the new geodetic reference system for the Brazilian Geodetic System (SGB) and the National Cartographic System (SCN) (IBGE, 2005). From January 2015, only SIRGAS 2000 will be official and all cartographic products are recommended to be referenced to this *Datum.*

exported to ArcMap (ArcGis 10.2).

In a new *Data Frame*, the orthorectified aerial photographs at a scale of 1:25,000 (EMPLASA-Projeto Mapeia São Paulo, 2011) and the control points (properties) were added. Thus, the 10 points were recognised (Quandro 1).

Points	Description	Latitude (Decimal)	Longitude (Decimal)	GMS- Latitude	G MS- Longitude
1	St Jesus	-22.331.708	-47.705.952	22°19'54.15"S	47°42'21.43"O
2	Sítio do Trevo	-22.323.761	-47.693.841	22°19'25.54"S	47°41'37.83"O
3	Good Hope	-22.316.747	-47.716.883	22°19'0.29"S	47°43'O.78"O
4	Chá caraSão Paul	-22.342.183	-47.680.247	22°20'31.86"S	47°4O'48.89"O
5	Maria's farm	-22.314.775	-47.701.613	22°18'53.19"S	47°42'5.81"O
6	Sítio Giobela	-22.333.594	-47.703.480	22°20'0.94"S	47°42'12.53"O
7	Old house	-22.316.944	-47.699.477	22°18'52.90"S	47°41'58.12"O
8	Old house	-22.314.455	-47.699.580	22°18'52.04"S	47°41'58.49"O
9	FormerFEPASA Station	-22.313.469	-47.699.266	22°18'48.49"S	47°41'57.36"O
10	Chapel of the Divine Saviour	-22.314.925	-47.701.152	22°18'53.73"S	47°42'4.15"O

Table 1: Control points Source: Personal archive, 2015.

It was therefore considered that on a scale of 1:10,000 the maximum acceptable deviation should be up to 3 metres. This means working with the Cartographic Accuracy Standard (PEC), Class A (0.5 mm on the map scale). Considering that, at a scale of 1:10,000, 1 mm on the scale means 10 metres from reality (terrain), so 0.5 mm means 5 metres from the reality of the terrain. Thus, the desired accuracy for georeferencing was achieved in relation to the Mean Squared Error (MSLE).

The delimitation of the area made it possible to extract the contour lines (with an equidistance of 5 metres) in polyline format from the topographic maps, with a table of attributes and their respective elevation values; the hydrography as a polyline; the points as points, with a table of attributes and their elevation values; and the roads as a polyline. The Clip tool was used to finalise by cutting out the elements that were outside the delimited area.

To build the clinographic map, we first had to generate the MNT or MDT[3] , using the 3D Analynst - Create Tin From Features tools in ArcMap (ArcGis 10.2), inserting the contour lines (elevation), datum points (cotado) and the basin boundary, automatically generating a terrain elevation model,

[3] Numerical Terrain Model or also known as Digital Terrain Model.

i.e. a thematic map of hypsometric classes (Altitude), which was reclassified into 8 classes. From this generated layer, we chose the Face slope with graduated colour ramp function, which allowed us to visualise the slope of the terrain in degrees.

This data was then reclassified into 6 categories in terms of degrees (1.14; 2.86; 8.50; 16.70; 24.22; 45), and these values were based on the percentages adopted (0-2%; 2-5%; 5-15%; 15-30%; 30-45%; 45%>), which were defined according to adaptations of the proposals by De Biasi (1989) and EMBRAPA (1995), in accordance with the specific characteristics of the Sapezeiro catchment area.

Slope class (EMBRAPA, 1995).			Slope class (adapted from De Biasi, 1989).		
Class	%	Relief (type)	Class	%	Definition
A	0-3	Plan	1	0-2	Strip with no restrictions on use, but with the possibility of flooding
B	3 - 8	Gentle waves	2	2-5	5% Urban-industrial boundary, used in urban planning carried out by IPT[4] and Emplasa[5]
C	8 - 13	Moderately undulating	3	5 - 15	15% Maximum limit on the use of mechanisation in agriculture.
D	13-20	Wavy	4	15-30	30% Limit defined by Federal legislation - Law 6766/79 - as the maximum limit for unrestricted urbanisation, whereby the parcel must be developed in accordance with specific requirements.
E	20-45	Strongly undulating	5	30-45	45% Forest Code, sets the maximum slope limit for clear-cutting and Law No. 12.651/12, of 25 May 2012 dictates that this strip can
F	>45	Steep	6	45 >	

[4] Technological Research Institute

[5] Metropolitan Planning Company of Greater São Paulo.

					only be explored with forest cover Forest Code, Art. 10 restricts the clearing of
					forests, with only the extraction of stumps being tolerated.

Table 2 - Slope classification
Source: Organisation, V. A. Souza, 2015.

In the clinographic map generated automatically in the ArcGis software, there were classification errors; in order to correct them, it was necessary to use the formula by De Biasi (1992, p. 48), which establishes the relationships between the classes chosen and the spacing between the contour lines and/or the gradient between the points on a map. As this is a 1:10,000 scale map, with an equidistance between the contour lines of 5 metres (n), shapefiles were created in Polygon format for all the slope classes for the appropriate correction.

$$D = \frac{n \times 100\%}{E}$$

D = Slope

E = Spacing or horizontal distance between the two consecutive curves.

n = Difference in level between the two points or the equidistance of the chart.

2.1.1 Geological and Pedological Chart

To draw up the thematic maps of the pedological and geological units, the Soil Map from the semi-detailed pedological survey of the state of São Paulo of the São Carlos grid SF23-Y-A-I, scale 1:100,000 (OLIVEIRA and PRADO, 1981) and the Geological Map, scale 1:50,000, obtained from the Geological Institute (IG, 1986) were imported into ArcGis, digitised in tiff format with a resolution of 300 dpi. The vectorisation of these maps enabled the extraction of the pedological and geological elements of the Córrego Sapezeiro catchment area, generating maps on a scale of 1:50,000.

2.1.2 Land Use and Occupation Chart

To draw up the Land Use and Occupation Chart, we used orthorectified aerial photographs from the Mapeia São Paulo Project, from 2010/2011, on a scale of 1:25,000 (EMPLASA, 2011), with the aim of analysing and understanding the problems and difficulties the region has been facing in the recent past, which limit agricultural activities and consequently the development of small rural properties.

The photographs requested that cover the study area have the following nomenclature/articulation (mapeiasp.sp.gov.br, 2015):

- SF-23-Y-A-I-4-NO
- SF-23-Y-A-I-3-NE

Thematic mapping was drawn up based on photointerpretation, identification and deduction of land use and occupation classes, which were later confirmed during fieldwork. Once the classes present in the study area had been identified, it was decided to use the Land Use and Land Cover System for Use with Remote Sensing Data (ANDERSON, 1979, P. 32) and the use of the colour code (USGS, Table 3), in which adaptations were made to the local reality.

Level I	Colours	Level II
1. urban or built-up land		Residential
		Commercial and services
		Industrial
		Transport, communication and utilities
		Industrial and commercial complexes
		Mixed urban or built-up land
		Miscellaneous urban or built-up land
2. Agricultural Land		Crop and pasture land
		Orchards, woodlands, vineyards, nurseries and ornamental horticulture areas.
		Confined breeding activity
		Other types of agricultural land
3. Pasture		Herbaceous pasture
		Grassland with shrubs and herbaceous vegetation
		Mixed pasture
4. Forest Land		Deciduous forest land
		Land of evergreen forest
		Mixed forest land
5. Water		**Watercourses and canals**
		Lagos
		Reservoirs
		Bays and estuaries
6. Wetland		Forested Wetland
		Unforested Wetland
7. Barren land		Salty and dry plains
		Beaches

		Other sandy areas
		Exposed rocks
		Open-cast mines, quarries and gravel pits.
		Transition areas
		Mixed land
8. Tundra		Shrub and mace tundra
		Herbaceous tundra
		Bare soil tundra
		Wet tundra
		Mixed tundra
9. Perennial Snow or Ice		Snowfields
		Perennial Glacier

Table 3 - Land Use and Land Cover System for Use with Remote Sensing Data and the use of the colour code (USGS)
Source: Anderson, 1979, p. 32.

With the adaptations made to this system shown in Table 2, the following land use and land cover system was generated for the watershed in question (Table 4):

Level I	Colours	Level II
1. urban or built-up land		Residential, commercial, services and rural property
2. Agricultural Land		Annual crop
		Citri culture
		Sugarcane
		Forestry
		Poultry farming
		Other confined and unconfined farming activities
3. Pasture		Pasture
		Dirty Pasture
4. Forest Land		Riparian and hillside forests
5. Water		Watercourses and canals, lakes and reservoirs
		Fish farming
6. Wetland		Wetland
7. Barren land		Open-cast mines, quarries and gravel pits.

Table 4 - Land Use and Land Cover System for the Córrego do Sapezeiro Watershed - Rio Claro (SP)
Source: Adapted from Anderson, 1979, p. 32-69. Organisation: V. A. Souza, 2015.

The digital aerial photographs provided by EMPLASA (2015) were imported into the ArcGis (10.2) software environment, where shapefiles of the classes previously defined in the photointerpretation activity were created.

The perimeters of the use and occupation classes were vectorised manually and then merged using the Merge tool and linked to a table of attributes in order to calculate the area values of each class in square kilometres.

The results have been structured into four parts, the first of which discusses the proposed theme, the multifunctionality of the rural landscape; the second part deals with the characterisation of the physical environment in the context of agricultural activity, highlighting the geomorphological, geological, pedological, climatic and land use aspects and natural cover of the study area; the third part provides information on the historical process of occupation and the socio-economic aspects of the Itapé rural neighbourhood; the fourth part concludes with an integrated analysis of the data, characterising the use and occupation of the Córrego Sapezeiro watershed, where the Itapé - Rio Claro (SP) rural neighbourhood is located.

CHAPTER 3

RESULTS AND DISCUSSION

3.1 Multifunctionality of the rural landscape

3.1.1 Discussing the landscape

The landscape is made up of a set of natural elements (formed by nature) and/or artificial elements (created by human beings). It is constantly changing and undergoing successive transformations, and will differ according to the perception of those who conceive it and the use that is made of it, the cultural, physical, economic and social context, adding to the appreciation and intentionality with which it is perceived. This means that the landscape is historically constructed and perceived by man. As such, perception is of great importance in defining the concept of landscape, which according to Melo (2005, p. 4),

> landscape is apprehended by taking its totality into account, in a holistic way. In this sense, the entire environment that surrounds man, whether physical, social or imaginary, influences his behaviour. Reality is interpreted and phenomena are observed as part of a larger, integral phenomenon, with the landscape being perceived by the individual not as a sum of objects next to each other, but simultaneously.

Therefore, the rural landscape is a social construction conditioned by the nature that surrounds it, and as Galvão (2010, p. 64) points out, three interconnected figures should be emphasised:

> The "resource" space, of the production systems and producers who participate in the landscape dynamics; the "frameworks of life" space, the scene of all the more or less dense modalities of so-called dispersed occupation; the "nature" space, in a more systemic vision that integrates resources, frameworks of life and all the biophysical functions of nature, as understood in its contrast with culture.

By defining functions for the landscape, goods and services are established that are supported by it, and that meet or are intended to meet the needs, demand and expectations of those who turn to them or who are part of them; some have their own needs.

market value, the so-called commodities, and other public goods and services, the amenities. According to PINTO-CORREA (2008, p. 67-71),

> functions are the goods and services supported by a landscape and which respond to the needs, demands and expectations of man, in the broadest sense, and are therefore valued by society [...] Some have a market value (agricultural production, forestry production), and correspond to amenities (commodities). Others are amenities, which correspond to public goods and services for which there is no market or the market does not function satisfactorily (recreation, nature conservation, identity, quality of life, preservation of environmental resources).

From the 1960s onwards, the introduction of new technologies and the homogenisation of rural space, especially with the "Green Revolution" model of agriculture, reduced the importance of the landscape to its productive function with the domination of the industrial sector, which brought new forms of production to the countryside, implying a new organisation of the landscape and, together, degradation due to indiscriminate use.

Thus, based on capitalist discourse, the dichotomy between urban and rural is fostered, with the former seen as "new", a place of "progress", and the rural as "old" and "backwards". This concept of the rural as equivalent to agricultural space, in a productivist view, would require a process of modernisation, which would involve developing the means of agricultural production, with the illusory idea that this would result in social welfare, in a socio-economic view. According to Graziano (1998, p. 43-81),

> the urban-rural dichotomy therefore sought to represent the social classes that contributed to the emergence of capitalism or opposed it in 17th century Europe, rather than a geographical division. It was from then on that the "urban" came to be identified with the "new", with the capitalist "progress" of the factories; and the rural - or the "landowning class", with the "old" (i.e. the old social order in force) and with "backwardness" in the sense that they sought to impede the progress of social forces.

Modernisation has profoundly altered the agrarian structure, contrary to what was proposed, driving small producers who were unable to adapt to the new

production techniques to indebtedness and even the sale of the property, and consequently to poverty.

The main changes in these transformations can be listed: the rural exodus, the decline and ageing of the population, the insolvency of traditional farming methods, the outsourcing of the economy, the lack of jobs and appropriate public policies. On the other hand, however, there has been a return to looking at the rural landscape from perspectives other than production.

The demand for the countryside, whether as a residential alternative (weekends and holidays) or as a tourist alternative, is associated with ideologies that promote the flight from the city to the countryside (the latter understood as a refuge from the troubled environment that surrounds the city). The countryside and its landscape are reminiscent, above all, of tranquillity, roots, "authenticity", due to its peculiar characteristics, the natural environment and freedom, harmony and quality of life.

In this way, certain areas may lose their ability to compete in terms of production, but they will develop other functions such as: support, recreation, nature conservation, environmental quality, aesthetics (especially for tourists, who will use the landscape for consumption) and the preservation of cultural and architectural heritage. These functions will be more highly valued, because not all rural spaces will be productive spaces, but they can develop other functions linked to the post-productive demand for rural space.

With this, we consider the interrelationship between four main points, which are essential for sustainable development, respecting human conditions and needs and natural conditions, which Cristovão (2013) highlights as points that should not be considered in isolation, because one is strictly related to the other and all together are fundamental for human development; these are: 1) Agriculture; 2) Environment; 3) Health; 4) Food.

Therefore, we must analyse the landscape from the perspective of the variety of products that rural activity is capable of providing, which go beyond basic needs, thus not restricting itself to the local area, but involving, both directly and indirectly, society in general. The idea of the multifunctionality of the rural landscape, the term of which, according to Mesquita (2005), has been used by the WTO[6] (2001), has three meanings:

1. Externalities.
2. Public property.
3. Non-economic objectives.

[6] World Trade Organisation

Incorporating these elements mentioned above into two propositions:

1. There are multiple commodities and other products that are produced jointly by agriculture;

2. Some products have the characteristics of externalities or public goods, and the market for these products either doesn't exist or functions poorly.

Thus, the WTO (2001, apud PINTO-CORREA, 2008, p. 67-71) defined multifunctionality by noting that:

> Agriculture is multifunctional when it has one or more functions that go beyond its primary function of producing food and fibre. Agricultural activities shape the landscape, provide environmental benefits such as soil conservation, sustainable management of renewable natural resources and the preservation of biodiversity, and contribute to the socio-economic viability of many rural areas.

The discussion on the multifunctionality of agriculture gained prominence in 1999 at the Seattle Conference, during the WTO negotiations, where negotiations on the reform of important international trade issues should have begun, with agriculture and services being the main focus. However, there was no consensus among WTO member countries, leading to failure and criticism of the WTO's structure, which was seen as undemocratic and lacking in transparency.

However, the debate on the multifunctionality of the rural landscape did not begin with these discussions in Seattle, but rather at ECO-92[7] , held in Rio de Janeiro in 1992, recognising the aspect of multifunctionality specifically in relation to food security and sustainable development.

As early as 1998, the OECD[7] [8] , an organisation that brings together the richest countries, pointed out that in addition to their primary function (producing fibre and food), agricultural activities can shape the landscape, provide environmental benefits and contribute to socio-economic viability in various areas of the countryside. This debate doesn't end here, as there are several other contributions that define the role of multifunctionality as a way of expanding the possibilities of rural existence in an environment that is increasingly merging with the urban, minimising its importance.

Thus, it can be summarised in the words of (CARVALHO, 2004; MACHADO, 2008;

---

[7] United Nations Conference on Environment and Development, also known as ECO-92, held in 1992.

[8] Organisation for Economic Co-operation and Development

CARNEIRO, 2003; FROEHLICH, 2005; N. GUIOMAR, 2007; KAGEYAMA, 2008) that agriculture is multifunctional when it has one or more functions that go beyond the primary role of producing food and fibres.

Rural multifunctionality will take into account not only income generation, but also food quality, food security, the sustainable use of natural resources, with sustainable production methods, taking into account the health and well-being of animals, the maintenance of biodiversity, the preservation of ecosystems, landscapes, heritage and cultural values, thus managing the space while respecting human, animal and natural conditions. Cristovão (2013) highlights three strategic points of the multifunctionality of rural production, be it tangible (material) or intangible (immaterial), which are: deepening (agroecology method, environmental and cultural functions), broadening (diversification) and refounding (gaining efficiency, plurativity and reorganising production). As well as valuing the peculiarities of rural and agricultural life, the concept of multifunctionality encompasses those that are not just production and private goods, and criticises the unsustainability of agricultural production and its products of dubious quality, thus seeking to break away from the sectoral approach, which restricts its role, and start to consider the heterogeneity of rural life and its benefits. To this end, important resources are considered, such as territorial capital and social capital, which, according to Cristovão (2013), include:

- territorial capital: financial, natural, cultural and political;
- social capital: people trained to manage the territory.

Working with the countryside as a space for living, working and leisure, and not just for agricultural production, opens up new possibilities for families to stay, guarantees food security, environmental protection, the preservation of tangible and intangible cultural heritage and, consequently, its development, conceiving development in its essence, not as a synonym for economic growth (measurable by growth in GNP and GDP), but as a process of overcoming social problems, in which a society becomes fairer and, of course, legitimate for its members. Emphasising the benefits that this rational and sustainable use can bring us. Galvão (2010, p. 72) emphasises that,

> the multifunctionality of agriculture results in a varied set of benefits: benefits for the rural community (maintenance of cultural traditions, local markets for

> products and production factors, **family farming** businesses...); **environmental benefits (hydrological regulation, soil protection...) and other benefits on a** national **scale** (conservation of the **countryside, food security...) (GALVÃO, 2010, p.72).**

Therefore, multifunctionality will play an important role in expanding the possibilities of rural existence, emphasising its variability, its heterogeneity, that is, its multiple functions that are not necessarily limited to its materialities (production and environmental preservation), but also in its immaterial characteristics, which are not restricted only to the population that lives there, but to society as a whole.

As we have seen, the discussion around the multifunctionality of the countryside has come to be addressed due to the need to expand the functions of the rural environment, since with the indiscriminate use of land in order to increase production, the homogenisation of the landscape, the introduction of technologies aimed at accumulation without taking into account the people who lived there, were part of it and depended on it, rural landscapes have been degraded.

The conditions of neglect and marginalisation that led to the rural exodus, due to the precarious quality of life, working conditions and the major environmental and cultural impacts that involved and still involve production maximisation practices, make us think about new strategies for the conservation and maintenance of our ecosystems and ways of solving these problems of the system.

With proper soil management, combined with land use that respects its limitations and considers its potential, we can be much more productive and for a longer period of time. What we have today is the increasingly intensive use of natural resources, based on the development discourse. In this context, soil degradation is one of the main problems affecting agriculture as a result of this unbridled use, and it is necessary to adopt alternatives to minimise these problems, reconciling conservation practices with other functions that rural landscapes can offer us.

Multifunctionality therefore has the requirement of demonstrating the great role that the countryside plays in society, both in enabling and expanding its functions, as a supplier of food and fibre, but also of amenities, which are non-agricultural in nature, without excluding them, and can be reconciled to provide greater support for permanence.

Thus, by offering "new functions" to the countryside, without disregarding the current context of the demand for the countryside as a means of making up for the shortcomings of the troubled urban environment (escape from the hustle and bustle of the city) and, of course, the population's need to find their roots, preserve rural cultural traditions, conserve ecosystems, natural resources, landscapes and maintain the territory, especially in terms of infrastructure, rural tourism was thought of as an alternative to achieve these goals.

Rural tourism today plays a fundamental role in land management, due to its ability to stimulate the internal development potential of a given location, not limited to agricultural activities, but encompassing modalities that do not exclude them, but complement them, linking them to the bucolic rural landscape, its historical, cultural and environmental heritage, both tangible and intangible.

3.2 Characterisation of the physical environment in the context of agricultural activity

The study area, the rural Itapé neighbourhood, is located in the watershed of the Sapezeiro stream, a tributary of the Cabeça river, which in turn is located in the watershed of the Corumbataí river. The Itapé neighbourhood is located to the northwest of the municipality of Rio Claro - SP, covering an area of approximately 23 Km^2 , with the coordinates 22° 18'S and 22° 21'S and 47° 39'W and 47° 45'W as references (Figure 2). The main access to the neighbourhood is via the Washington Luiz Highway SP-310, passing through a stretch of unpaved side road, totalling 24 km from the city of Rio Claro to the rural neighbourhood.

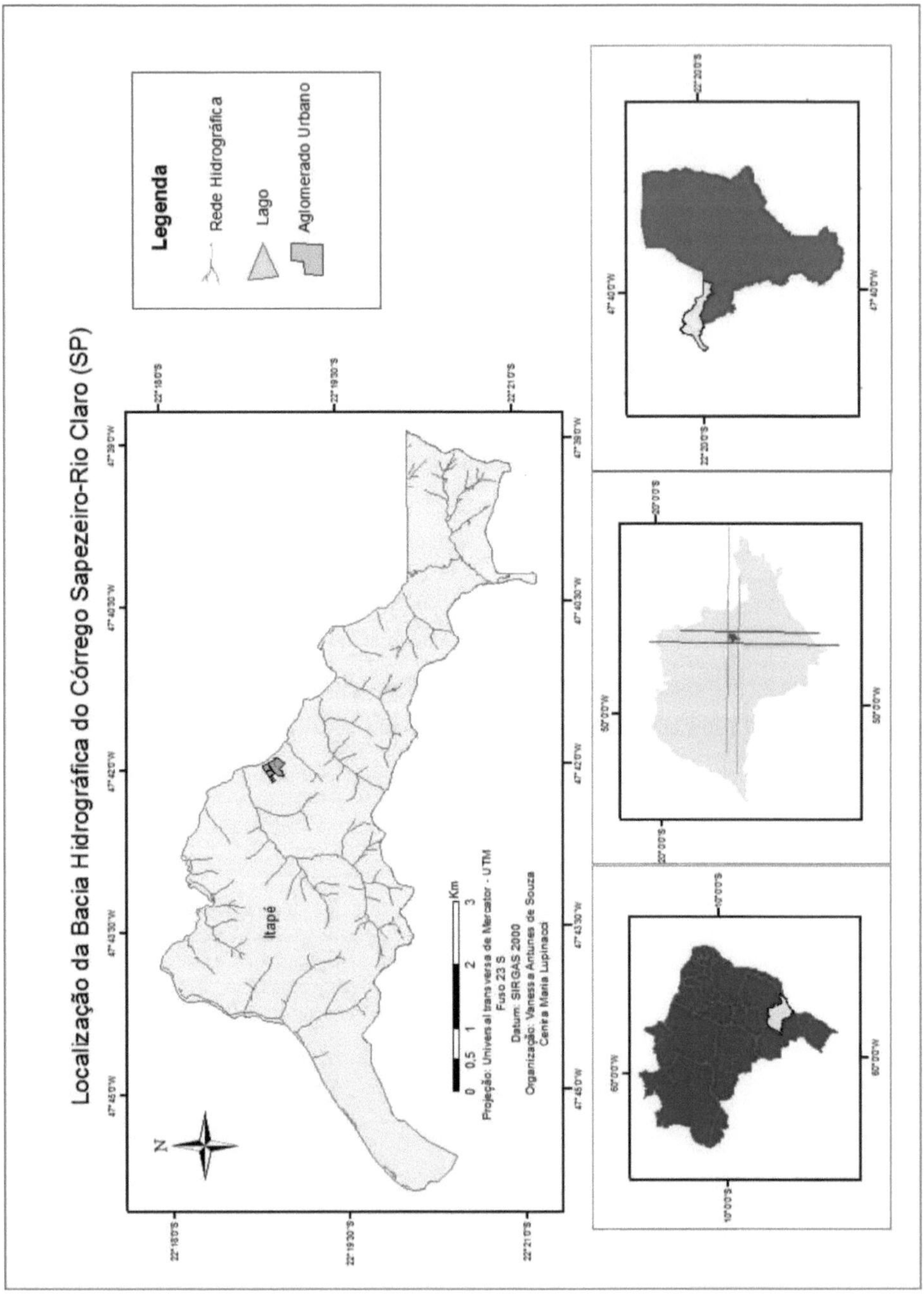

Figure 3 - Location of the Córrego Sapezeiro watershed/ Itapé rural neighbourhood Source: Organização,V. A. Souza, 2015.

3.2.1 Geomorphological aspects

According to Penteado (1968), the main relief of the Rio Claro region is made up of tabular hills with gently convex slopes and gently sloping plateaus, arranged between 550 and 650 m, delimiting the west, north-west and north, lining up with the sandstone-basalt escarpments of the Serra de Itaqueri, Morro Guarita, Serra do Cuscuzeiro and Morro Grande as a summit surface between 950-1000 m.

These sandstones capped by basaltic flows are from the Botucatu Formation, which form these escarpments and hills that break up the monotony of the landscape, standing out from the friable sediments of the underlying Estrada Nova Formation (PENTEADO, 1968).

According to the IPT (1981), the basaltic rocks, which are more resistant than the underlying ones, allow for a process of differential erosion in which the incision of a subsequent drainage perpendicular to the direction in which the layers fall exposes lithological alternations of varying resistance to denudational processes.

The unprotected eastern sectors of the spills are quickly eroded, forming a topographically lowered strip, the Paulista Peripheral Depression between the edges of the spills to the west and the crystalline outcrop to the east. To the west, the basaltic packages between the Palaeozoic and Mesozoic depositions made it possible for them to be better preserved, forming high topographic levels, the Planalto Ocidental Paulista (Western Plateau of São Paulo), which falls gently to the west, following the slope of the layers. Between the Peripheral Depression and the Western Plateau, the limits of the basaltic flows support steep cuesta fronts (figure 4), whose rapid retreat leaves behind a relatively rugged relief, largely marked by witness hills and asymmetrical valleys (AB'SABER, 1956; CASSETI, 2001).

In an intermediate position, between this ridge line and the lower general interplanaltic surface, the Serra Santana stands out, as well as the isolated testimonies in front of it, capped with sandy sediments recognised as "modern". This prominent surface within the regional morphological framework was named by Penteado (1968) as the Intermediate surface because it lies between the summit surface and the Interplains.

The descent from the edge of the western plateau into the Rio Claro basin is made through clear structural and erosive steps, giving this depressed compartment the characteristic shape of an amphitheatre (PENTEADO, 1968). In terms of lithology, the sediments that occupy the most space in the area studied are sandstones from the Pirambóia formation (aeolian sediments), which will be described in more detail below (Figure 3).

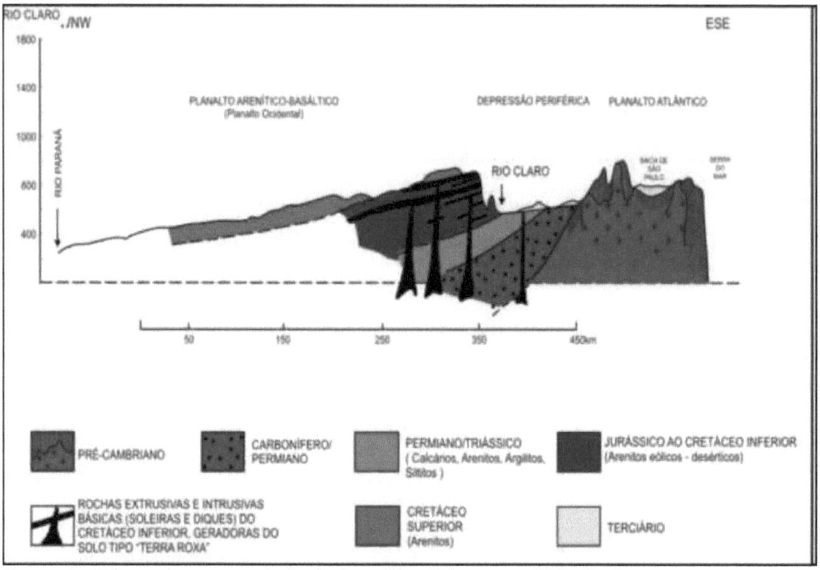

Figure 4 - Schematic geomorphological profile of the state of São Paulo
Source: AB'SABER, 1956.

The route between the city of Rio Claro and the rural Itapé neighbourhood takes place in the midst of these geomorphological elements, i.e. in a transitional relief landscape between the Peripheral Depression and the Western Plateau. According to the IPT (1981), the city of Rio Claro is located 600 metres above sea level, in the Middle Tietê Zone[9] of the Paulista Peripheral Depression, in the transition between forms of hilly relief, with broad hills and relief of elongated hills with spikes.

3.2.2 Lithostratigraphic units in the study area

The study area is geologically located in the São Paulo sector of the northeastern flank of the Paraná Sedimentary Basin (ZAINE, 1994, p. 13), represented by sedimentary and igneous

9 DEFFONTAINE, 1935; MONBEIG, 1949; ALMEIDA, 1964 distinguish three distinct natural sub-regions: 1) Mogi-Guaçu Zone; 2) Middle Tietê Zone; 3) Parapanema Zone. According to the authors, Rio Claro is located in the Middle Tietê Zone.

(volcanic) rocks from the Palaeozoic (Corumbataí Formation of the Passa Dois Group), Mesozoic (Pirambóia Formation of the São Bento Group) and Cenozoic (with Recent Tertiary and Quaternary Deposits and the Rio Claro Formation) eras. The following sections will describe in more detail the most significant units found in the Córrego Sapezeiro catchment area (Figure 5).

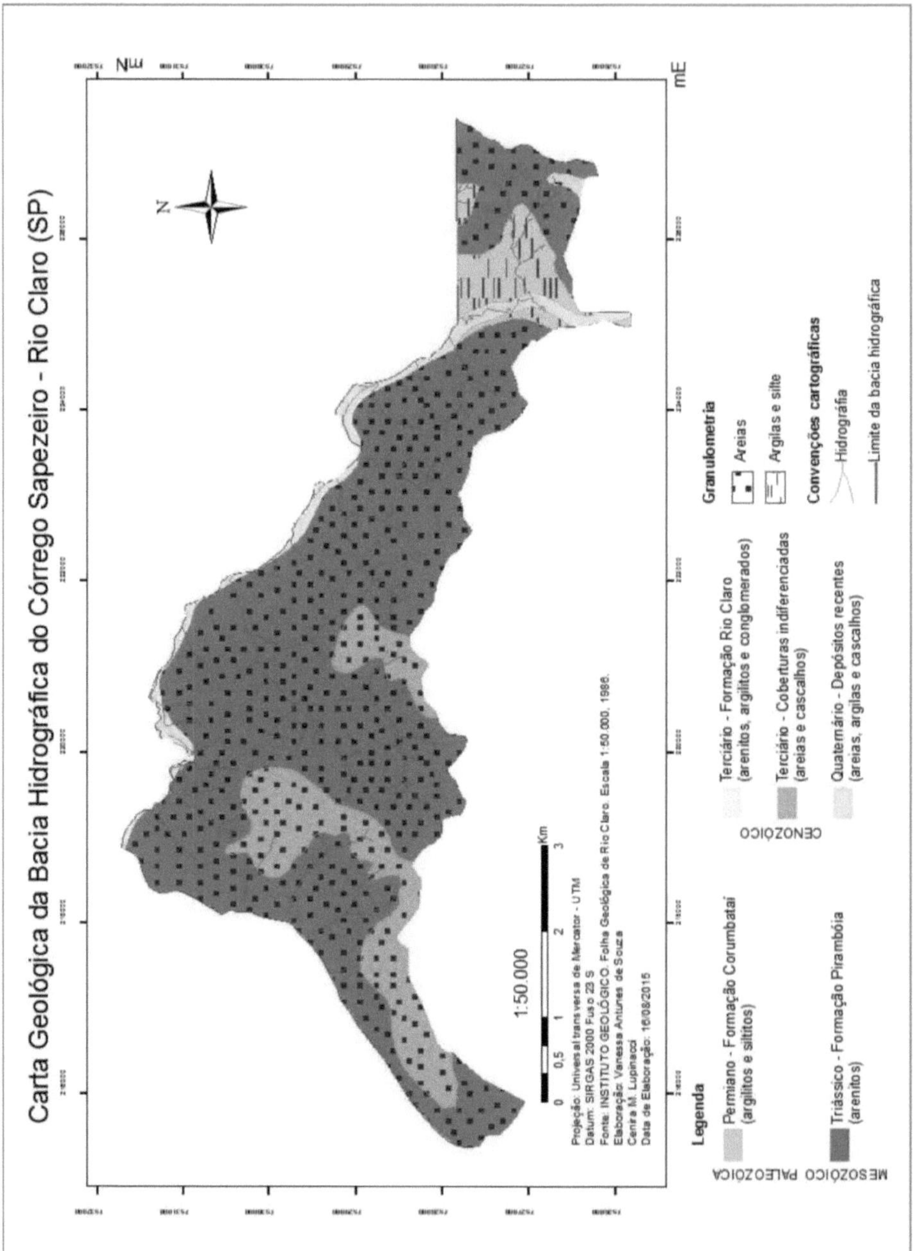

Figure 5 - Geological map of the Córrego Sapezeiro basin - Rio Claro (SP) Source: Organisation, V. A. Souza, 2015.

3.2.2.1 Corumbataí Formation

The Corumbataí Formation belongs to the Passa Dois Group of the Upper Permian period, of the Palaeozoic Era; it is made up of older and less expressive deposits in the study area, located more in the southeastern part of the Córrego Sapezeiro basin. The deposits of this Formation are attributed to the marine plains and include grey, purplish or reddish claystones, shales and siltstones with intercalations of carbonate banks, silexites and layers of fine sandstones (IPT, 1981, p.61).

It shows plane-parallel stratification, low-angle cross-stratification, rhythmic stratification, undulating marks, drying fractures and also intra-formational breccias. This formation is in discordant contact with the Pirambóia Formation, which predominates in the study area (IPT, 1981, p. 61).

Fossils can be seen in the Corumbataí Formation, and some of the rocks may be oolitic, rich in pelecypod shells or silified stromatolites. According to Landim (1970, apud IPT, 1981, p.61), this formation is relatively rich in fossils, with many sedimentary structures indicating that they arose in coastal conditions dominated by tidal currents and others in environments such as lakes.

Shneider (1974, apud IPT, 1981, p.62) identifies the lower part of the formation as a shallow-water marine environment in colder climatic conditions. The upper portion, on the other hand, is the result of deposition in shallow waters under oxidising climatic conditions, with the influence of tides. According to Gama Jr. (1979, apud IPT, 1981, p. 62) the Corumbataí Formation is related to a deltaic system constituting a tidal plain that developed in the state of São Paulo. This formation is the main supplier of raw materials for the region's potteries and for the industries in São Paulo.

Santa Gertrudes Ceramic Pole. Some mining areas in this lithological unit can be found in the study area.

3.2.1.1 Pirambóia Formation

The Pirambóia Formation, which belongs to the São Bento Group from the Triassic period, Mesozoic Era, is made up of deposits that predominate in the study area, covering almost the entire river basin. It is characterised by its deep sandy layers, most commonly red, whose granulation varies from medium to fine, with clay formation in the lower part of the formation, where thick, conglomeratic sandstones occur. The deposits of this formation are attributed to the humid, oxidising continental environment, predominantly fluvial, meandering and floodplain, with scattered small lagoons (IPT, 1981).

It shows plane-parallel stratification emphasised by the alternation of laminae more or less rich in clay and silt or cross, medium or large tangential stratification, showing wave or current marks. According to Fúlfaro et. al. (1980, apud IPT, 1981) these marks are related to the lithological variations observed in the formation, which characterise the overflow channel facies. This formation is known for the importance of its rocks as Reservoir Rock, being the main groundwater reservoir of the Guarani Aquifer in the state of São Paulo, as well as providing raw materials for the glass and foundry industries (CAETANO-CHANG; WU, 2003 p. 33).

In this way, we realise the importance of thinking about appropriate use and management, respecting the capacity and limits of the formations and the soils that originate from them, as they are very sandy, friable soils that are susceptible to erosion, such as the Red-Yellow Argissolo in a large part of the study area.

3.2.1.2 Rio Claro Training

The deposits of the Rio Claro Formation belong to the Cenozoic Era, from the Tertiary Period, and are found in the study area as a small patch at the south-eastern end of the river basin. It is characterised by poorly sorted, reddish-yellow, friable sandstone sediments with cross-stratification, conglomeratic levels, with clay laminae up to 30 metres thick. There are a large number of quartz pebbles and quartzites at its base.

According to the IPT (1981, p.92), the chronology of this formation is uncertain, but many authors believe it to be of recent geological age, such as Bjornberg et. al. (1964), who gave it a "fairly young" geological age, Neocenezoic, due to the fact that there are plant fossils that give it that age.

According to Bjornberg & Landim (1963, apud IPT, 1981, p.63) the sediments of this formation were deposited in a semi-arid climate, during the development of pediments; while Fulfaro and Suguio (1974) defend the idea that they originated in a fluvial environment (small lagoons) under tectonic imposition, in a semi-arid climate.

There is great economic interest in this formation due to the extraction of its sands, which are used in construction and in the glass and foundry industry, as well as the extraction of underground water (mineral), which also gives it economic importance (CAETANO-CHANG; WU, 2003).

3.2.1.3 Recent deposits

The deposits found are from the Cenozoic Era, the Tertiary and Quaternary Periods, and these are distributed throughout the study area as small isolated patches in the form of undifferentiated Tertiary coverings (sands and gravels) and Quaternary colluvial deposits.

3.2.3 Pedological units in the study area

The soils that cover the study area are predominantly characterised by dystrophic red-yellow Argisols (PVA), moderate A, sandy/medium texture, and in the western part of the catchment area; with less spatial expression of dystrophic red-yellow Latosols, moderate A, medium texture. There are small patches of deep quartz sand (Quartzarenic Neosols), alkalic, moderate A (AQ); remnants of lithic soils (Litholic Neosols); and purple latosol in the western part of the watershed, where it is on a steep slope (EMBRAPA, 1994), at an altitude of between 700-900 metres (Figure 6).

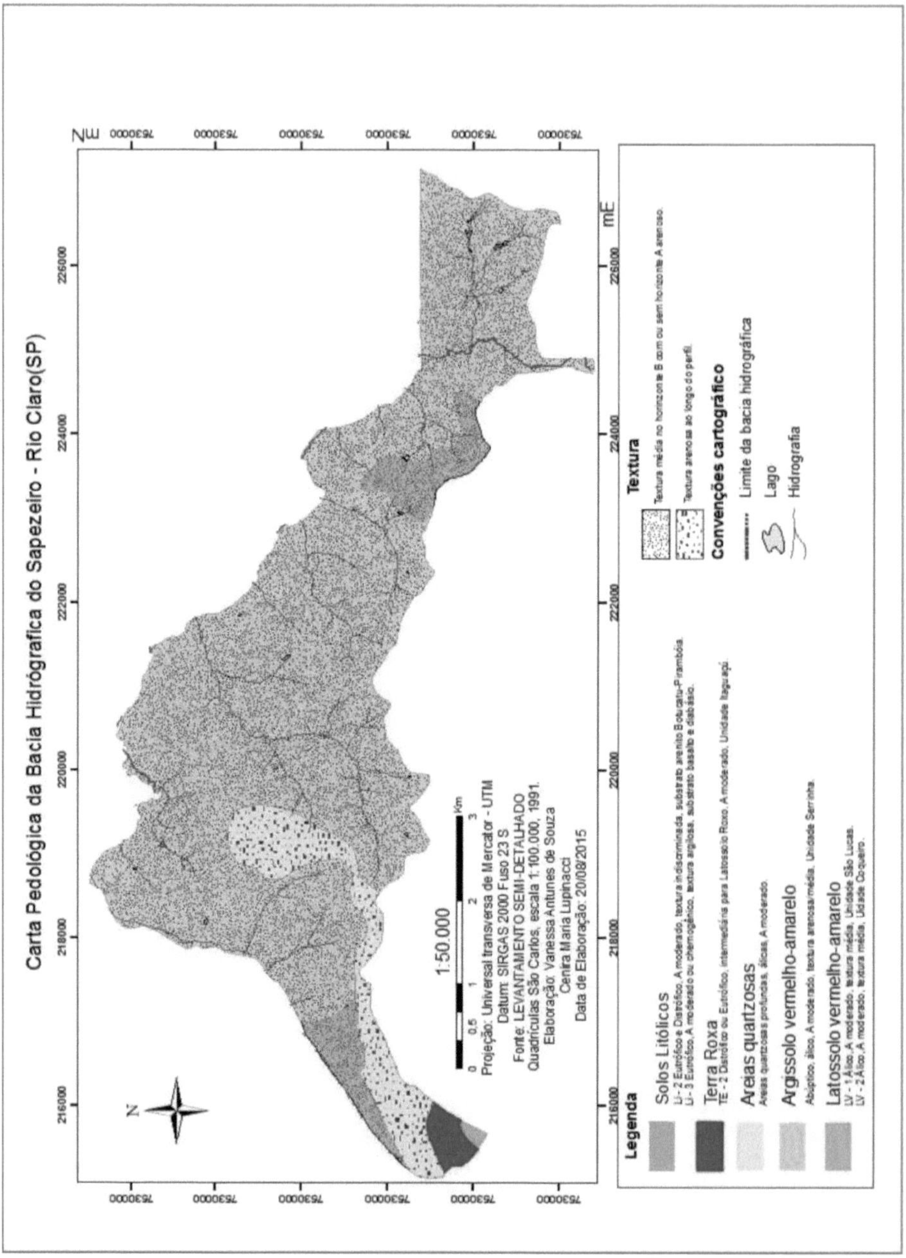

Figure 6 - Pedological map of the Córrego Sapezeiro basin - Rio Claro (SP) Source: Oraganisation, V. A. Souza, 2015.

We chose to describe the pedological characteristics according to the classification of the Brazilian Soil Classification System (EMBRAPA, 1999, 2005), which defines:

3.2.3.1 Argisols

Soils consisting of mineral material with low-activity or high-activity clay combined with low base saturation or an alkaline character and a textural B horizon (Bt) in the layer following the A or E horizons.

Suborder recorded in the study area:

3.2.3.1.1 Red-Yellow Argisol

Argisols with a 5YR matrix or more red and more yellow than 2.5YR in most of the B horizon, including BA. Dystrophic in cases where the B horizon has base saturation <50% and aluminium saturation <50%.

The increase in clay at depth and a cation exchange capacity of less than 27 mol/Kg are the main diagnostic attributes valid for Argissolos. Red-yellow Argisols have a low percentage of iron oxides and a lower phosphorus fixation capacity.

Due to the greater textural relationship between the A or E horizons and the textural B horizon, it has little cohesion on the surface and little permeability in subsequent horizons, generating greater susceptibility to erosion. Erodibility is generally very high in soils with abrupt textural changes, in which case intensive erosion control practices are required when they are under agricultural management (IAC, 2000).

In soils with abrupt textural changes, it is common to see an aeration zone between the E horizon and the top of the Bt horizon during the rainy season due to the lower hydraulic conductivity at the top of the Bt horizon. When the terrain is not very steep, in rainy periods, sloe red-yellow Argisols are more difficult to drain internally, which is detrimental to certain crops that are more sensitive to waterlogged soils. In summer, however, this characteristic can be beneficial for plants, as these soils can have greater water availability (IAC, 1999 p. 11).

Argissolos, on the other hand, have a medium or sandy texture on the surface and low activity of the clay fraction, and are easily prepared for planting. These soils are deep, with no physical impediments to the penetration of the root system up to a depth of 200 cm. According to Oliveira (1999, p.14), the original vegetation was tropical sub-deciduous forest, but it is currently being taken over by sugar cane monoculture and pasture.

3.2.3.2 Latosols

Soils made up of very altered mineral material, with a brassolic B horizon, immediately below any type of A horizon, within 200 cm of the soil surface or within 300 cm if the A horizon is more than 150 cm thick (IAC, 1999, p. 50-52).

Suborder recorded in the study area:

3.2.3.2.1 Red-yellow latosols

These are soils with a 5YR matrix or more red and more yellow than 2YR in most of the first 100 cm of the B horizon.

Latosols are soils with good physical properties and are generally located on slopes that are favourable to the use of agricultural machinery, as they have little movement and slopes of less than 5%. Due to their high friability, they are easily prepared for cultivation. They therefore qualify as soils suitable for extensive agriculture. However, they have limitations when they are dystrophic, as they have low nutrient availability and Al^3 + (exchangeable aluminium) toxicity, making them very acidic; in these cases, they need corrections and fertilisation for good productivity.

They have good internal drainage, even with a clay texture. Medium-textured latosols have a high sand content, giving them rapid permeability and poor water retention due to the predominance of macropores. These soils dry out quickly after rain or irrigation, and are more prone to water stress in summer and more susceptible to erosion in winter.

According to Manfredini et al. (1984, apud OLIVEIRA, 1999 p.52) in medium-textured latosols there is a 50% increase in water storage capacity and a decrease in the hydraulic conductivity of saturated soils, as the average diameter of the sand particles decreases. Because of their high permeability and high absorption capacity, they are classified as filter soils.

Thus, in these soils there is a high possibility of contamination of aquifers by toxic material deposited in them. On the other hand, this porosity, coupled with the fact that they are quite deep, creates suitable conditions for good root development. According to Oliveira (1999, p.112) they were originally vegetated by Cerradão, but are currently being taken over by pastures.

3.2.3.2.2 Purple Latosol (Terra Roxa)

In the region you can see remnants of this type of soil on the western edge, with eutrophic (high fertility) and dystrophic (low fertility) characteristics, moderate A, very clayey texture, intermediate to Latossolo Roxo. They are soils resulting from the decomposition of diabase with probable relocation (IPT, 1981, p.55).

As already mentioned, these soils have good physical properties due to the fact that they are quite clayey and have exceptional porosity, with good internal drainage. According to Lombardi Neto (1975 apud IPT, 1981, p. 52), Latosols have a good tolerance to erosion loss and a low erosion ratio. The low activity of the clays in these soils gives them little expansiveness and contractibility, and in this case their limitation is particularly related to the low availability of nutrients in dystrophic soils.

3.2.3.3 Neosols

As the name suggests, these are new, undeveloped soils that do not have a defined diagnostic B horizon, with individualised A horizons followed by C and R. This soil has strong characteristics of the original material, due to its lack of development, either because of the low activity of pedogenetic processes, or because of the inherent characteristics of the original rock, such as resistance to weathering, which can prevent and limit the evolution of soils.

Neosols are characterised by their shallow depth and low moisture retention, limiting their use for agriculture. They are soils well supplied with nutrients, in some cases, such as those in the study area where they derive from basic rocks such as basalt and diabase.

3.2.3.3.1 Litholic soils

The litholic soils (Neossolos Litólicos) found in the study area are eutrophic and dystrophic, moderate A, indiscriminate texture, Botucatu-Pirambóia sandstone substrate. And eutrophic, moderate A or chernozemic, clay texture, basalt and diabase substrate.

The limitations to its use are more related to its shallow depth, the presence of rocks (original material) and steep slopes (EMBRAPA, 1994), with a slope of over 45% at an altitude of between 800-900 metres.

3.2.3.3.2 Quartz Sands (Quartzarenic Neosols)

The Quartzarenic Neosols (previous classification Quartzose Sands) present in the study area are deep, A moderate quartzose sands from the sandstones of the Pirambóia Formation. They are devoid of weatherable primary minerals, have very low colloidal activity and also low water and nutrient retention capacity, due to their sandy characteristics which give them low adhesion and cohesion, making them very susceptible to erosion.

Quartz sands are poor in nutrients, very porous and highly permeable. According to the IPT (1981, p. 77), their lack of nutrients makes it essential to apply inputs in order to achieve satisfactory production. However, due to its low buffering power, it requires applications of inputs in instalments in order to minimise losses and avoid saturation of the soil's sorption complex. In these soils, there is a high possibility of contamination of aquifers by toxic material deposited in them.

3.2. 4Climate conditions

According to the Koppen Climate Classification, the study region has a rainy and moderately warm temperate climate (C), in the subdivision (CWa), hot and rainy summer, with the warmest month having an average temperature greater than 22°C (a), with a dry winter (W) and mild temperatures (MENDONÇA, 2007, p.119-121) .

Brino (1973, apud CEAPLA, 2015) considers that the region of the Corumbataí River basin, where the Córrego Sapezeiro watershed is located, falls within the sub-tropical climate, with annual temperatures of 20.9°C at the confluence of the Passa Cinco and Corumbataí rivers and in the areas close to São Pedro - SP.

In January, there is a variation between averages of 20°C and 24°C, while in July the averages are 14.9° and 17.1°C.

Graph 1 shows the air temperatures (average minimum, average maximum) and average rainfall (mm) in the municipality of Rio Claro - SP, according to data from the Centre for Meteorological and Climatic Research Applied to Agriculture - CEPAGri (2015).

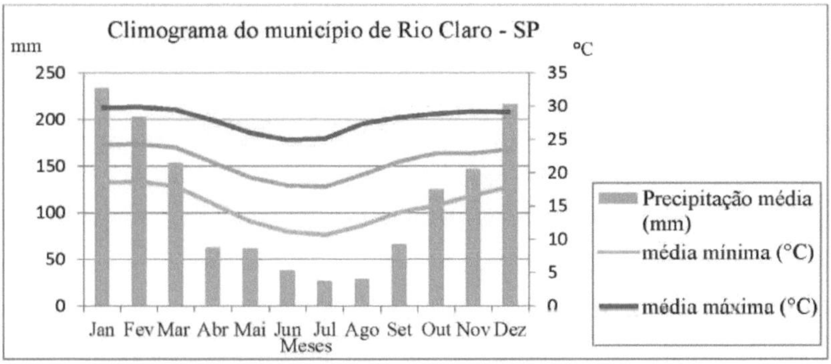

Graph 1 - Climatogram of the municipality of Rio Claro (SP)
Source: CEPAGri/UNICAMP - Climate of São Paulo municipalities Available at: < http://www.cpa.unicamp.br/outras-informacoes/clima_muni_494.html Accessed on: 20/07/2016.
Organisation: Vanessa Antunes de Souza.

Rio Claro (SP) is markedly sub-tropical (wet and dry), with a cool winter (monthly average of around 17°C) and a hot summer (monthly average of 24°C). It can be seen that as the temperature rises in summer, rainfall increases significantly, i.e. from September to March. The annual temperature range is considered very low.

3.2.5 Land use and natural vegetation cover

The original predominant vegetation in the Paulista Peripheral Depression was Cerrado, Cerradão and Atlantic Forest. In the case of the Cerrado, the predominant forest transitions according to the morphology and pedology of the relief. At the top of the cuestas, the sandstone-basalt formation allows for greater infiltration of rainwater, thus forming sheets of water that rise up, giving rise to many springs.

Delimiting the western, north-western and northern regions of the municipality of Rio Claro, there are some elements that define and individualise themselves, both in terms of the shapes themselves and in terms of the sculpting processes of the relief, which create contrasts, evidenced through the natural landscape and human occupation itself.

Thus, the region's landscapes are characterised by the Tropical Slope Forest on the front of the cuesta, remnants of cerrados on its reverse side, the morphological aspects of the hilly terrain with broad tabular hills with gently convex slopes, low sloping plateaus with a great predominance of sugar cane monoculture and elongated hills with spikes.

The combination of different vegetation formations and the morphology of the terrain creates a rich landscape marked by the presence of a high concentration of biodiversity in the areas of dense vegetation, a strong bucolic landscape that is interrupted by the razing of sugar cane monoculture activities.

The type of vegetation that can be seen on the route from the city of Rio Claro to the Itapé neighbourhood is basically the result of anthropogenic activities, such as sugar cane, orange, pine and eucalyptus orchards. The area's original vegetation is no longer visible, as it has been practically devastated by sugar cane monoculture activities. It only has remnants in areas of environmental preservation, forest reserves, areas where there are tourist activities, among others.

3.3 Occupation history and socio-economic characterisation of the rural neighbourhood of Itapé

The beginning of the occupation of the Sapezeiro stream basin, where the Itapé neighbourhood is located, is closely associated with the history of the Rio Claro Railway Company railway, later acquired by Companhia Paulista de Estrada de Ferro (later FEPASA[10]), which connected the municipality of Rio Claro to São Carlos and Araraquara. As well as other stations between Rio Claro and Itirapina. The Itapé station (Figure 7) was built and inaugurated in 1916, however, at the time it was without tracks, limiting its functions for almost a year to administrative activities.

[10] Ferrovia Paulista S.A.

Figura 7 - Itapé railway station (1918)
Source: F. Peres, available at: < http: //www.estacoesferroviarias.com .br/i/itape.htm> Accessed on: 20/03/2016.

The installation of Itapé station was due to the ratification and widening of the gauge built by Paulista (broad gauge, Figure 8), which transformed the stretch of the Serra de Corumbataí into the "Analândia Branch" (FEPASA, Fixed Installations Report, 1986).

Figura 8 - Fragment of the map of the lines of the Companhia Paulista de Estrada de Ferro (CPEF)
Source: R. P. Corrêa, available at: <http://efpaulista.blogspot.com.br/2014_03_01_ archive.html>
Accessed on: 20/03/2016.

The station had five lines, and according to old residents of the neighbourhood, at loading times more than 100 people worked there.

carried various products that were produced on site, such as potatoes, rice, coffee, fruit, sand and other goods. The flow of people at the station was significant: in addition to the workers, there were vendors, shoppers, onlookers who watched the trains arrive, among others.

In 1980, the Rio Claro - Nova station was opened, also called Guanabara, as it was located in the neighbourhood of the same name. This newly opened Santa Gertrudes-Itirapina bypass, which passes outside the city to the west, replaced the line that passed through the Serra de Corumbataí. The Itapé station was deactivated that year and the tracks were removed. After the 1980s, the station fell into disrepair, along with the neighbourhood's original buildings, and from then on it was gradually demolished house by house. Today there are only two houses (Figures 9 and 10) and the old station building (Figure 11) of these original buildings in the neighbourhood.

Figure 9 - Old house (original building in the neighbourhood) Source: Fieldwork, 2016.

Figure 10 - Old house (original building in the neighbourhood) Source: Fieldwork, 2016.

Figure 11 - Old railway station building (FEPASA)
Source: Fieldwork, 2016.

According to FEPASA's report in 1986, the station was already in ruins, totally deteriorated. They report that even without the tracks, the station lasted for some time (FEPASA, Relatório de Instalações Fixas, 1986).

In 1996, the Rio Claro town hall renovated the building to make it fit for use, and it is currently used as a health centre for the rural district of Itapé. Few of the neighbourhood's residents know about the historical context of that time. According to residents who inherited the properties, their parents say that when the station was active, productivity was very good, and many people farmed in the area, like Mr Antônio, who has lived there for 57 years, is a cattle farmer, president of the football club and owner of the only commercial establishment in the neighbourhood. He says

> families used to be united, we used to get together at dances here in the region, our parents used to farm, everything was abundant, we had opportunities, today we're discriminated against, and the authorities disregard us and the oldest families have either left or are leaving, almost nothing is produced here any more... (verbal information, 2015) .[11]

You can see the indignation in their words and that history is being lost and families are no longer united as they used to be, as many who live in the neighbourhood today are new residents or heirs and are unaware of the local traditions.

In this part of the work, we will look at the socio-economic aspects of the neighbourhood's population, trying to identify their economic, cultural, business and educational potential and their expectations. The data was collected through fieldwork observations, recorded in a notebook and the application of forms with closed and open questions.

During the course of the fieldwork, we came across empty, closed properties, either because they were abandoned or because they were only used at weekends or for parties and events. Therefore, of the properties visited, information was collected on the spot from 33 properties.

The questionnaire data showed that Itapé's rural properties are predominantly small, i.e. most of them are rural houses, farms and ranches of up to a maximum of 50 hectares, i.e. approximately 4 fiscal modules[12] (Table 1).

Table 1 - Property size

Number of properties	Property size (ha)
1	0,072
1	0,14
1	0,16
5	2,42
2	3
1	3,5
1	3,63
1	4,84

[11] Interview given by Antônio (2015), during an interview in Itapé - Rio Claro, São Paulo, 2015.
[12] Fiscal Module of the Municipality of Rio Claro - SP/ Fiscal Module =14 ha (CATI, 2008).

1	6
4	7,26
4	9,68
1	12,82
4	14,52
1	16,94
1	24,9
1	31,4
1	33,8
2	38,7

Source: Personal archive, 2015.

Most of the local residents are caretakers, owners and their families (Table 2). The caretakers are responsible for the upkeep of the site, and many owners use the site as a leisure environment (weekend getaway), or as a home, but work in the city (Table 3), without producing anything on the property or producing through an employee or tenant. Pluriactivity can be seen in the area, which is other activities carried out by rural owners and their family members who live in rural areas, maintaining a productive link with agriculture and/or life in rural areas, while carrying out other non-agricultural economic activities.

> Pluriactivity is established as a social practice, resulting from the search for alternative ways to guarantee the reproduction of farming families, one of the mechanisms for reproduction, or even for expanding alternative sources of income; with the economic, social and cultural reach of pluriactivity, families living in rural areas are integrated into other occupational activities, in addition to agriculture (BAUMEL E BASSO, 2004, p. 139 apud PIRES; SPRICIGI, 2016, 20 p.1).

If we think of pluriactivity in the context of the multifunctionality of agriculture, non-agricultural activities should be part of an efficient and rational agriculture. However, what we are witnessing is pluriactivity seen as a reflection of the precariousness of agriculture, especially family farming.

> The plurative who dedicates himself to temporary wage labour can be seen as the result of a process of precariousness in family farming and the difficulties it faces in a reality dominated by the sugar cane and orange agro-industrial chains (BAUMEL E BASSO, 2004, p. 139 apud PIRES; SPRICIGI, 2016, 20 p.3).

Table 2 - Residents of rural properties

Residents	Properties (n°)
Lessee	1
Homemade	11
Employees	2

Tenants	1
Employees	1
Family members	1
No response	3
No-one	3
Owner and family	10

Source: Personal archive, 2015.

Table 3 - Owner's main activity

Owner's main activity	Property (n°)
Butcher	1
Farmer	4
Retired	4
Urban wage-earner	1
Social assistance	1
Autonomous	1
Poultry farmer	2
Trader	1
Clothing manufacturing	1
Housewife	1
Ceramist	1
Businessman (plastics factory)	1
Beautician	1
Photographer	1
Doctor	1
Micro-entrepreneur	1
I didn't want to answer	1
Livestock farmer	2
Bricklayer	2
Professor	3
Agricultural technician	2

Source: Personal archive, 2015.

Of the 33 interviewees, 15 are landowners (Graph 2), many of whom acquired their property by purchase (Graph 3), but claimed that their intention was to produce and that they had made some attempts to do so, but were very frustrated with the results, either due to low productivity, according to them, the region's soil conditions are very poor (sandy, acidic and poor soils), or the lack of public technical assistance aimed at small farmers, the lack of committed labour, and the difficulty of obtaining funding to buy agricultural inputs (mechanical, biological, mineral or chemical). The lack of security and isolation of the region, and consequently the theft of machinery, has also led some to abandon farming.

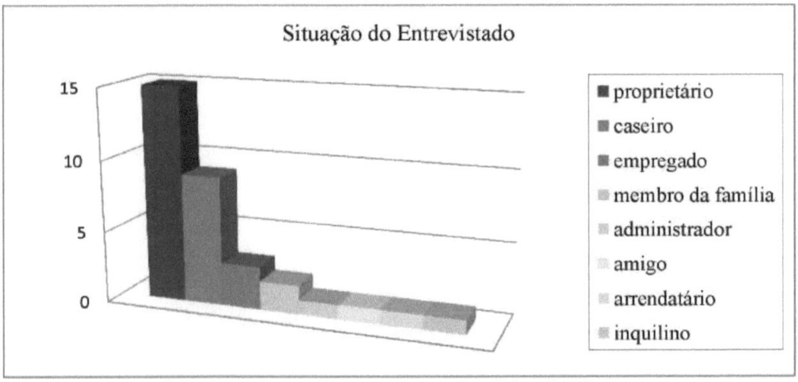

Graph 2 - Status of the interviewee Source: Personal archive, 2015.

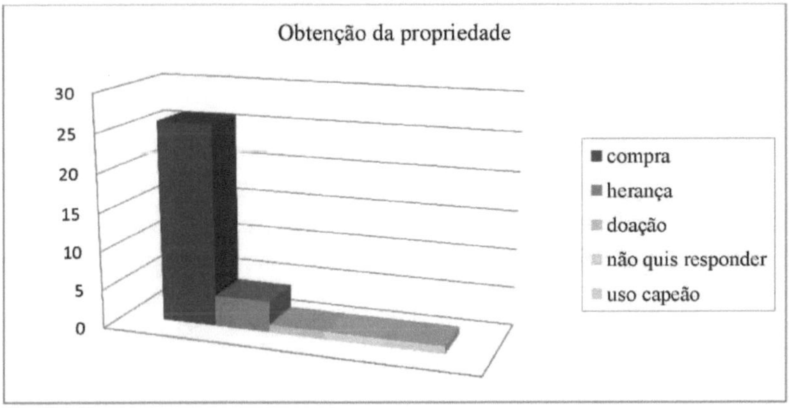

Graph 3 - Means of obtaining property
Source: Personal archive, 2015.

The 33 families interviewed very much regret the neighbourhood's lack of infrastructure and basic sanitation, as it has no piped water or sewage treatment, relying directly on wells, mines and springs (Table 4) and sewage collection is done by septic tank. The neighbourhood does not have a rubbish collection system, and everyone is responsible for the rubbish they produce, with interviewees claiming that they incinerate or bury their waste. With regard to energy resources (Table 5), they are supplied by the company Elektro, some have a generator but do not depend on it directly (for safety reasons).

Table 4 - Water supply

Water supply	Property (n°)
Mina	3
Spring	5
None	1
Country well	23
Artesian well	1

Rio	1

Source: Personal archive, 2015.

Table 5 - Energy resource

Energy resource	Property (n°)
Electricity (Elektro)	31
Generator	2
No response	1
None	1

Source: Personal archive, 2015.

All the interviewees have a mobile phone, and in some families more than one, as it is more viable for those who travel a lot during the day. As a result, many said that they considered their home telephone unnecessary. As for the internet, only 2 of the 33 interviewees have this service via a mobile device (Table 6).

Table 6 - Communication

Communication	Property (n°)
Mobile phone	33
Internet	2
I didn't want to answer	1
None	1
Landline	1

Source: Fieldwork (Questionnaire).

The lack of infrastructure in the neighbourhood causes indignation among its residents, given that it has been abandoned by the government. In the words of resident José M.:

> We are taxpayers, we pay taxes, but we are discriminated against by the local government, we don't have running water, we don't have rubbish collection, to catch the bus I have to go up the road and to do that I have to go over an improvised footbridge on the bridge that has been broken for more than three years and we don't know when it will be finished (verbal information, 2015) .[13]

The families have few expectations of continuing to live or produce in the area; many are thinking of pursuing other activities and think it would be preferable to rent their land and move to the city. On the other hand, the new families that are arriving are not farmers, but are settling there because of the services offered as caretakers of rural properties and/or as employees or because of the high prices of homes in the city, since Rio Claro is going through a period of

[13] Interview given by José (2015), during an interview in Itapé - Rio Claro, São Paulo, 2015.

intense property speculation.

Of the oldest families in the neighbourhood, few still live there (Table 7). Of these, 6 were interviewed, answering more specific questions about the neighbourhood. Those who remain are elderly pensioners who don't depend directly on income from agricultural activities,

These are families with roots in the land, who have always worked in agriculture, but who have given their children other opportunities to seek new forms of work in the city.

An example of one of the families mentioned above is the cattle farmer who inherited the property where he has lived for 57 years. Antônio said he wouldn't want to change the types of activities he carries out on his property, because that's what he's always done and the minimum the property produces is enough for his family's subsistence, providing a higher quality of life and food security that industrialised products and city life can't offer. His son went to study in the city, where he currently lives and works, but visits him and his wife at weekends. According to Antônio "my son is already raised and the only thing I've ever done in my life is work with cattle, I can't see myself doing anything else" (verbal information) .[14]

Young people from the area generally leave Itapé due to a lack of opportunities, going to work and study in the city, creating bonds with city habits. Many of these young people end up living in the urban area and visit their parents at weekends. Few stay in the countryside, remaining as commuters or self-employed in general services in the neighbourhood.

Table 7 - Length of ownership (years)

Time owned by family (years)	Property (n°)
1-10	12
11-20	2
21-50	12
51-80	7

Source: Personal archive, 2015.

Housekeepers are the main employees hired to maintain the properties visited, and they are also the ones hired to carry out agricultural activities, working alongside the family on the property (Graph 4) and producing for internal consumption and small-scale production.

[14] Interview given by Antônio (2015), during an interview in Itapé - Rio Claro, São Paulo, 2015.

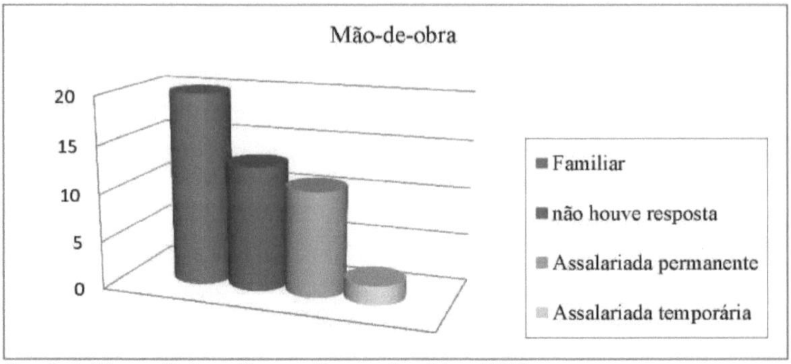

Graph 4 - Type of labour Source: Personal archive, 2015.

The main source of income comes from the salaries paid for maintaining the properties (Table 8), with a third of the population earning between 2 and 3 minimum wages from on-site activities. A third receive less and chose not to answer and a third receive more than 3 minimum wages, with income from outside activities, whose education is higher education (Table 3). Expenditure on food is low due to the subsistence agriculture produced, with the main crops and livestock being: vegetables, fruit, manioc, oranges, lemons, cattle (beef and dairy), chickens, ducks, quails and others (Graph 5).

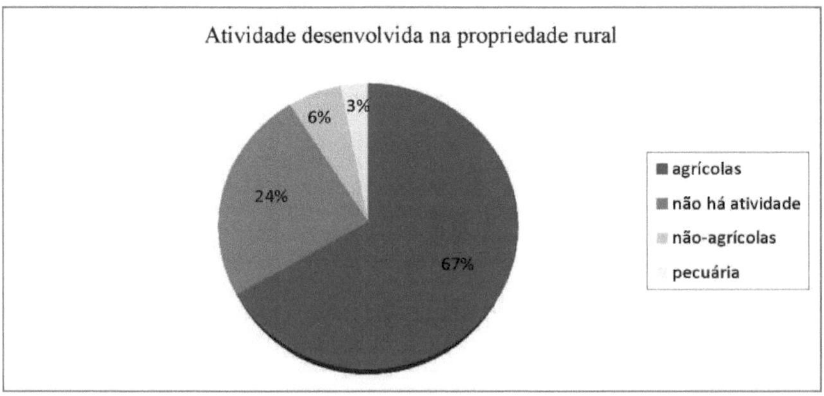

Graph 5 - Activity carried out on the property Source: Personal archive, 2015.

Table 8 - Family income (monthly total)

Family income (total monthly)	Property (n°)
No response	3
Up to 1 minimum wage	7
2 to 3 minimum wages	10
3 to 5 minimum wages	6

>5 minimum wages	7

Source: Personal archive, 2015.

The main agricultural activities carried out in the region, as already mentioned, are subsistence farming and the leasing of land for monoculture plantations of sugar cane, citrus (oranges) and cattle breeding. Those who said they had technical assistance totalled 21%, basically made up of the region's tenants, citrus growers and cattle farmers; the rest (79% of those interviewed) had no technical assistance (Table 9).

Table 9 - Technical assistance

Technical assistance	Technical assistance (cost)	Technical assistance Who?	Property (n°)
No	0,00	No response	26
Yes	0,00	Integrator	2
Yes	0,00	No response	3
Yes	70,00	Veterinarian	1
Yes	75,00	Veterinarian	1

Source: Personal archive, 2015.

The families' participation in the life of the community and the municipality is basically through the church, which according to them is the most active entity in the region, as it promotes religious and charitable events, and the football club organised by the residents of the neighbourhood, which serves as a means of local entertainment.

The community directs the search for solutions to local problems to public bodies, through a representative, who announces by word of mouth that meetings have been called (they take place in the old FEPASA building) with public representatives, but few attend. The population claims that their demands are disregarded, demanding other measures such as seeking out the media (EPTV, the local affiliate of Rede Globo) to express local problems.

Among the public policies and programmes aimed at family farming, the Itapé residents interviewed said they were unaware of them and that they were not effective in the area. They recognise the disregard and discrimination on the part of the public authorities towards the neighbourhood, both in terms of solving infrastructure problems and in terms of presenting appropriate public policies to solve rural difficulties, either because they are not part of Rio Claro's electorate or because some of them are voters in Itirapina or Corumbataí, neighbouring municipalities.

3.3.1 Itapé between resistance and neglect

Here we will analyse the impasses to rural development in the Itapé neighbourhood, and one of the problems blocking the region's development is related to public policies and infrastructure. The neighbourhood's main access is via the Washington Luís SP-310 highway, where it is necessary to take a diversion because the road has a toll, thus increasing the distance of the journey and/or the costs, whether of fuel or the toll for those who choose to pay.

According to the Casa da Agricultura de Rio Claro (2009, apud Melo, 2009 p. 9), in 2009 the rural neighbourhood of Itapé had more than 300 inhabitants, with around 80 properties, many of which are currently abandoned. The main access roads are not paved, are bumpy and have narrow stretches that make access difficult and are not maintained. Although there is a lot of criticism about the potholes that form in these places, there are residents who advocate "controlled" maintenance, without the need for paving, just maintenance based on materials such as earth and gravel.

On 29 and 30 January 2005, extremely high rainfall levels, recorded by CEAPLA[15] , i.e. 185.2 millimetres of rain hit the region and the floodgates of the Corumbataí Power Station broke, causing a lot of damage in the municipality of Rio Claro. Among the bridges that suffered damage, the bridge over the Cabeça River, which connects Itapé to the Washington Luiz Highway, collapsed in 2011 after rainy days, directly and indirectly affecting the population.

Residents' access to the town was made more difficult and they had to take an alternative route of more than 10 kilometres, which was much longer and the road was in poor condition, often being unfeasible, especially on rainy days. In 2012, the owners, tired of demanding a new bridge, organised themselves and took a palliative measure. In this way, the population organised to make repairs to the bridge, which had been closed since the event, in order to facilitate access to the neighbourhood for residents and visitors, and also to make it possible to hold the Feast of the Divine Saviour, the community's patron saint. The proceeds from the celebration, which included the sale of sweets, snacks and raffle tickets, were used to renovate the local chapel (Figure 12), as it was in terrible condition, as were other structures in the neighbourhood, such as the Basic Health Unit (USB).

[15] Centre for Environmental Analysis and Planning

Figure 12 - Chapel of the Divine Saviour
Source: Fieldwork, 2015.

As far as health is concerned, medical care in the area is only available on Saturdays from 7am to 11am at the UBS (Figure 13), whose operation and service is limited, subjecting the population to great risks in cases of urgency and emergencies, given that the quickest access to the town is via the Washington Luís motorway (SP-310).

Figure 13 - Former railway station (Current UBS) Source: Fieldwork, 2015.

The region has high rates of violence, and there is no police station in the immediate vicinity, leaving the population vulnerable to the most adverse situations. Residents are very sorry that all the properties in the area have already been robbed or stolen. And not just because of the robberies, but also because of the neglect and lack of police procedures, including forensics.

The problems don't stop there: the population complains about the lack of a school for their children, most of whom go to the school in the neighbouring municipality of Corumbataí, which offers more support to the residents of Itapé than the Rio Claro City Hall, a questionable fact, since the attributes are collected by the public authorities of Rio Claro - SP, which should therefore be in charge of the local improvements.

Public transport is also precarious, making it difficult for children to go to school and for the population to travel to central areas, and even school transport is provided by Corumbataí. The population regrets that even the receipt of mail is poor.

The construction of a sports court on the site left residents frustrated, as they had hoped for other more necessary actions. In April 2013, construction began on a new bridge under the supervision of the Rio Claro Municipal Works Department and the Piracicaba company Bema.

To carry out the work, the city council invested R$693,375.52 thousand. According to councillor Juninho da Padaria, who was on site at the start of the work, it was expected to last up to six months (a fact that didn't materialise) "The work could last up to six months and, during this period, the diversions for cars is around 10 kilometres long (verbal information)[16] " and he emphasised that, once the inconvenience was over, the residents of Itapé would have a definitive solution with regard to the bridge.However, after the installation of three beams made of USI-SAC-50 steel, weighing more than 24 tonnes, a job done with cranes and support equipment, the work was interrupted. Over the beams, concrete slabs were to be built to support the bridge deck, more than 25 metres long and seven metres and forty centimetres wide (Figure 14). The company Bema, responsible for the work, claimed that it had stopped the work because the municipality was in debt to the tune of 300,000 reais. They were therefore waiting for new negotiations to return.

[16] Interview given by Juninho da Padaria (2014), EPTV Central - Rio Claro, São Paulo, 2014.

Figure 14 - USI-SAC-50 steel beams
Source: Fieldwork, 2014.

In January 2014, in a note sent to the Jornal Cidade column, the Rio Claro City Hall press office revealed that it had renegotiated the debt with the company responsible for building the bridge in Itapé (Figure 15), finalising the work in July 2015.

Figure 15 - New Bridge
Source: Fieldwork, 2016.

There is therefore a great need to take swift action to ensure the permanence and socio-economic development of this population, which lives in and/or derives its livelihood from this region. Public authorities need to collaborate with efficient public policies, investing in infrastructure and basic sanitation so that they can develop and have the conditions to do so, benefiting the local community. And by opening up new possibilities, whether these are developed to complement or add value to existing products or as the main source of income, managed by the family itself and in accordance with natural, cultural and social potential.

3.4 Characterisation of the use and occupation of the Córrego do Hidrográfica Basin

Sapezeiro

The research region is located in the Preservation and Sustainable Use Macrozone, defined by the 2015 Master Plan (Figure 16), and is characterised by the area's environmental fragility, the presence of Environmental Protection Areas (APAs) and the importance of the availability of Rio Claro's natural resources. This area was determined according to the homogeneity of environmental characteristics and the degree of urbanisation, thus implying the form of land use and occupation planning.

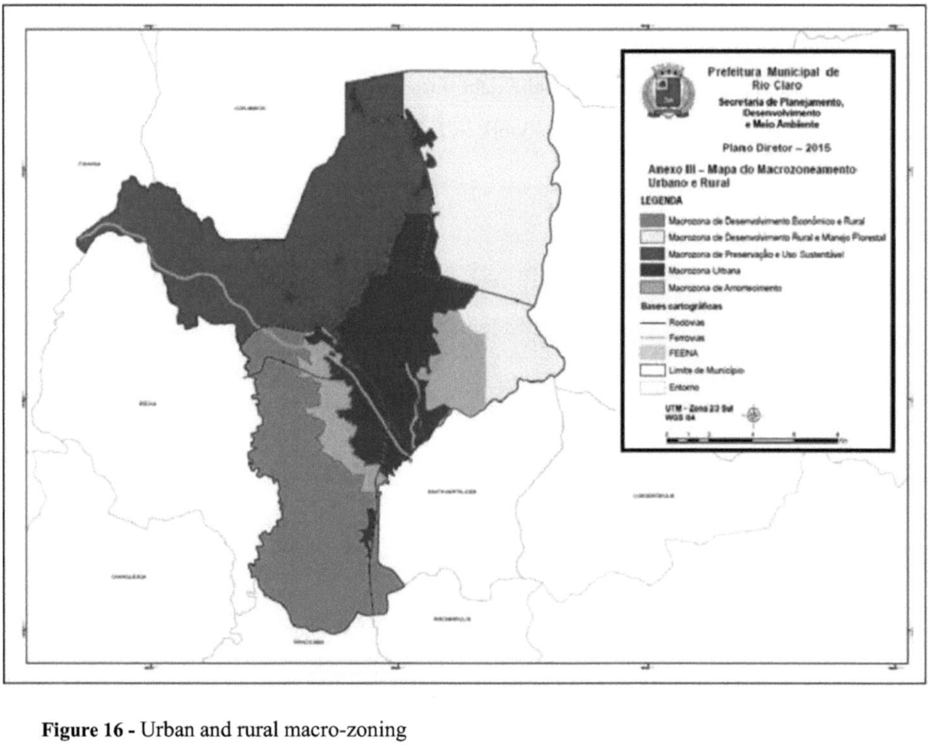

Figure 16 - Urban and rural macro-zoning
Source: Master Plan - Rio Claro City Hall; Planning Department, Development and Environment, 2015. Available at: <http://www.ríoclaro.sp.gov.br/pd/arquivos/pd2015/PD Rio Claro 2015 finalOLD.pdf<
Access:13/04/2016.

Land use in the study area is predominantly characterised by pasture (clean and dirty), which occupies 43% of the catchment area, 32% of which is clean pasture and 11% dirty pasture.

There are also significant areas of isolated woodland throughout the catchment area, totalling

23% of the area, especially in Permanent Preservation Areas (APPs) and Legal Reserves (RLs); these are found in areas of steep slopes, ecological corridors and scattered patches, and on the banks of river channels (riparian woodlands).

Sugarcane was also found in the study area with a significant share of 22% of the total area. The other classes found in the catchment area, such as forestry - 2%, poultry farming - 0.5%, fish farming - 0.5%, citrus farming - 0.6%, mining - 0.4%, annual crops - 1%, built-up area - 2% and wetlands - 5%, were recorded to a lesser extent, together accounting for 13% of the total area (Figure 16).

Graphs 6 and 7 show the distribution of the classes of types of use and occupation in relation to the total area of the watershed in percentage (Graph 6) and by area in square kilometres (Graph 7), where the rural neighbourhood[17] Itapé is located, which is also considered an isolated urban[18] , according to SEPLADEMA (2015).

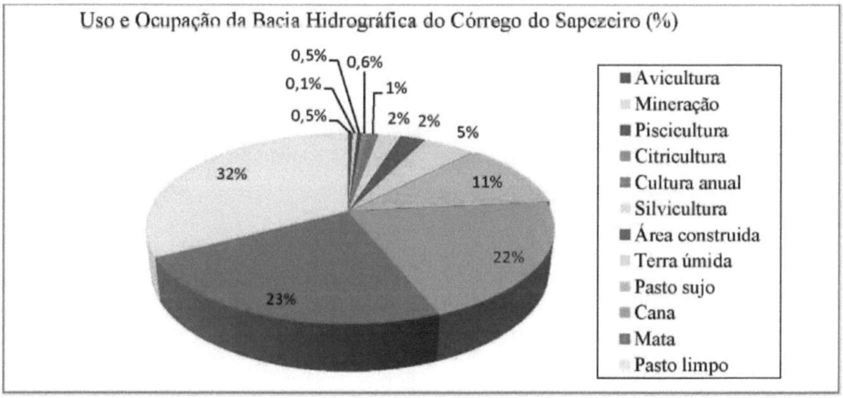

Graph 6 - Distribution of land use and occupation classes in relation to the area of the Córrego do Sapezeiro Watershed - Rio Claro (SP) - Area (percentage).
Source: Personal archive, 2015.

[17] The Department of Agriculture, Supply and Forestry defines Itapé as a rural neighbourhood.
[18] The Department of Planning, Development and Environment (SEPLADEMA) defines Itapé as an isolated urban centre.

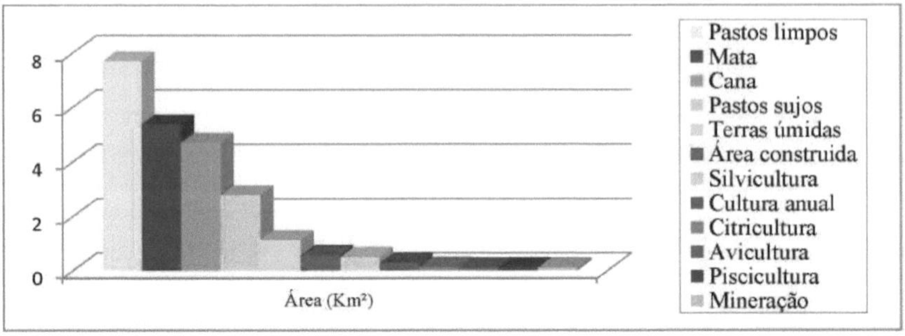

Graph 7 - Land use and occupation in the Córrego do Sapezeiro Hydrographic Basin - Rio Claro (SP) / Area (Km).[2]
Source: Personal archive, 2015.

By analysing the aforementioned data and the Use and Occupation Chart for the Córrego Sapezeiro catchment area (Appendix A), it can be seen how the land in the study area has been used, predominantly for pasture, woodland and sugar cane. Other types of activities such as fish farming, poultry farming, forestry, citrus farming and annual crops occur in a more sparse and less expressive way across the landscape, distributed throughout the catchment area as small plots, activities aimed at self-consumption and supplementing income, with the main income coming from
from other activities both in the city and from renting out land for monoculture sugarcane plantations and livestock farming.

So, by analysing the landscape of the Itapé neighbourhood from the perspective of multifunctionality, we can see that the variety of agricultural products that rural activity in the area is capable of offering may not be satisfactory at the level of large-scale market production, but rather for the subsistence of the families living there.

With these "new functions" of the countryside in mind, we analysed the multifunctionality of the Itapé neighbourhood, considering the countryside not only as a space for generating income, but also food quality, food security, the sustainable use of natural resources, thinking about sustainable production methods, taking into account the health and well-being of animals and local families, the maintenance of biodiversity, the preservation of ecosystems, landscapes, heritage and cultural values, thus managing the space and respecting human, animal and natural conditions.

We will therefore present the Use and Occupation Map of the Córrego do Sapezeiro

Watershed (Figure 17) to visualise and understand the spatial distribution of use and occupation in the watershed.

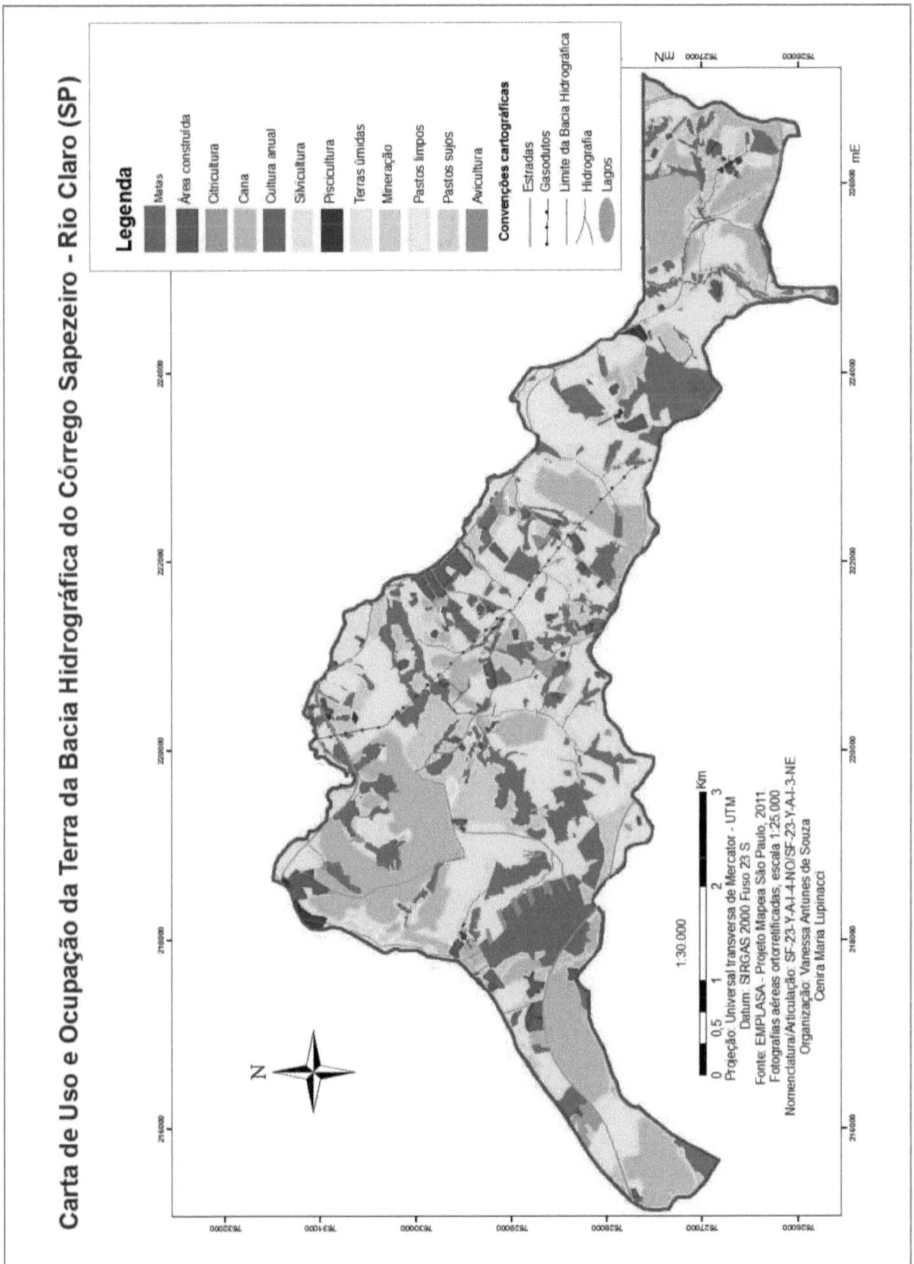

Figure 17 - Use and Occupation Map of the Córrego Sapezeiro Hydrographic Basin - Rio Claro (SP)
Source: Organisation, V. A. Souza, 2015.

3.5 Analysing the physical attributes of the landscape and the use and occupation of the Córrego do Sapezeiro Watershed

In order to analyse the watershed's physical attributes and the ways in which its use and occupation have been triggered, the study area was compartmentalised into sectors (Figure 18) to facilitate understanding and analysis, but it should be stressed that this fragmentation is purely of an analytical nature.

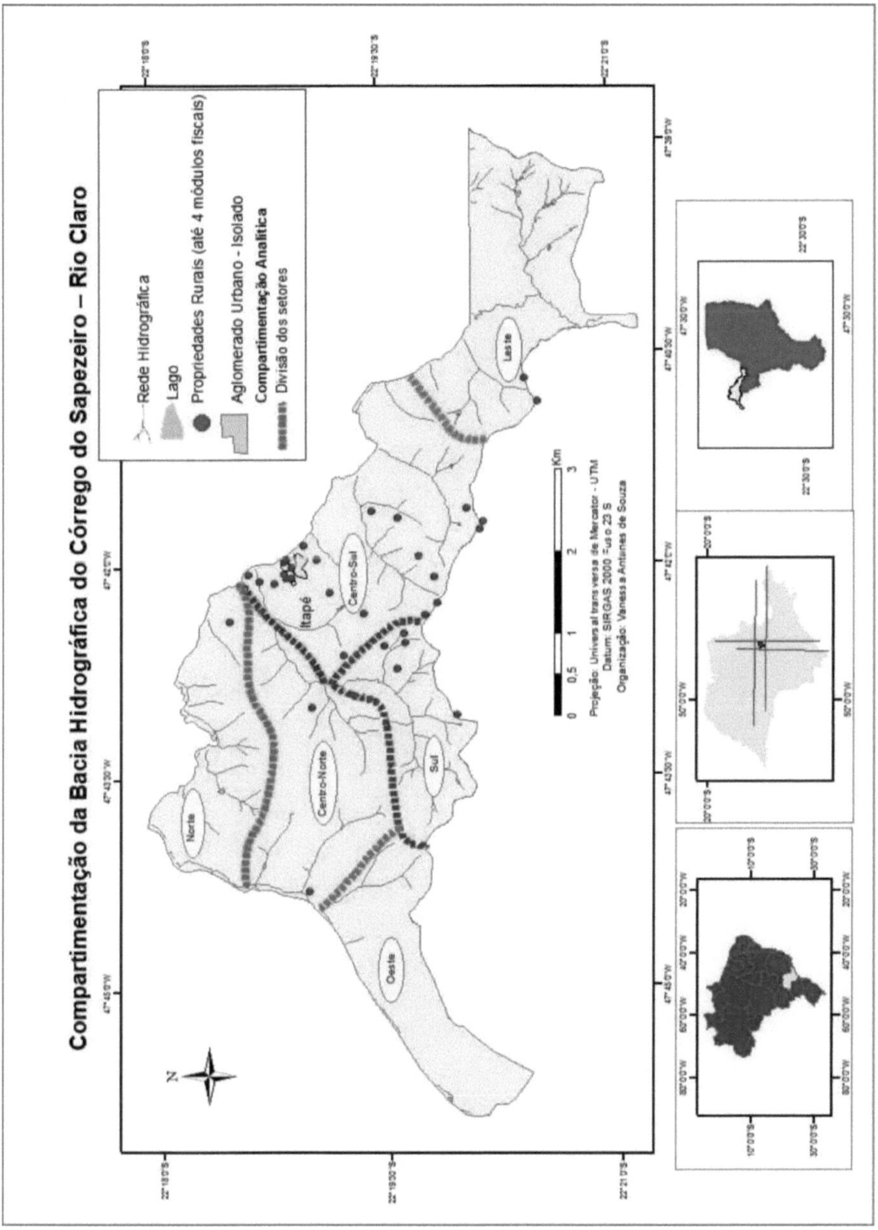

Figure 18 - Compartmentalisation of the Sapezeiro Stream Basin - Rio Claro (SP) Source: Organisation, V. A. Souza, 2016.

In addition, the Clinographic (Figure 19) and Hypsometric (Figure 20) Charts of the Sapezeiro Stream Basin are presented below, which made it possible to analyse the altitude in metres and the slope in percentages of these sectors, allowing us to have a better understanding of the local relief and equate the problems that occur in the analysed area.

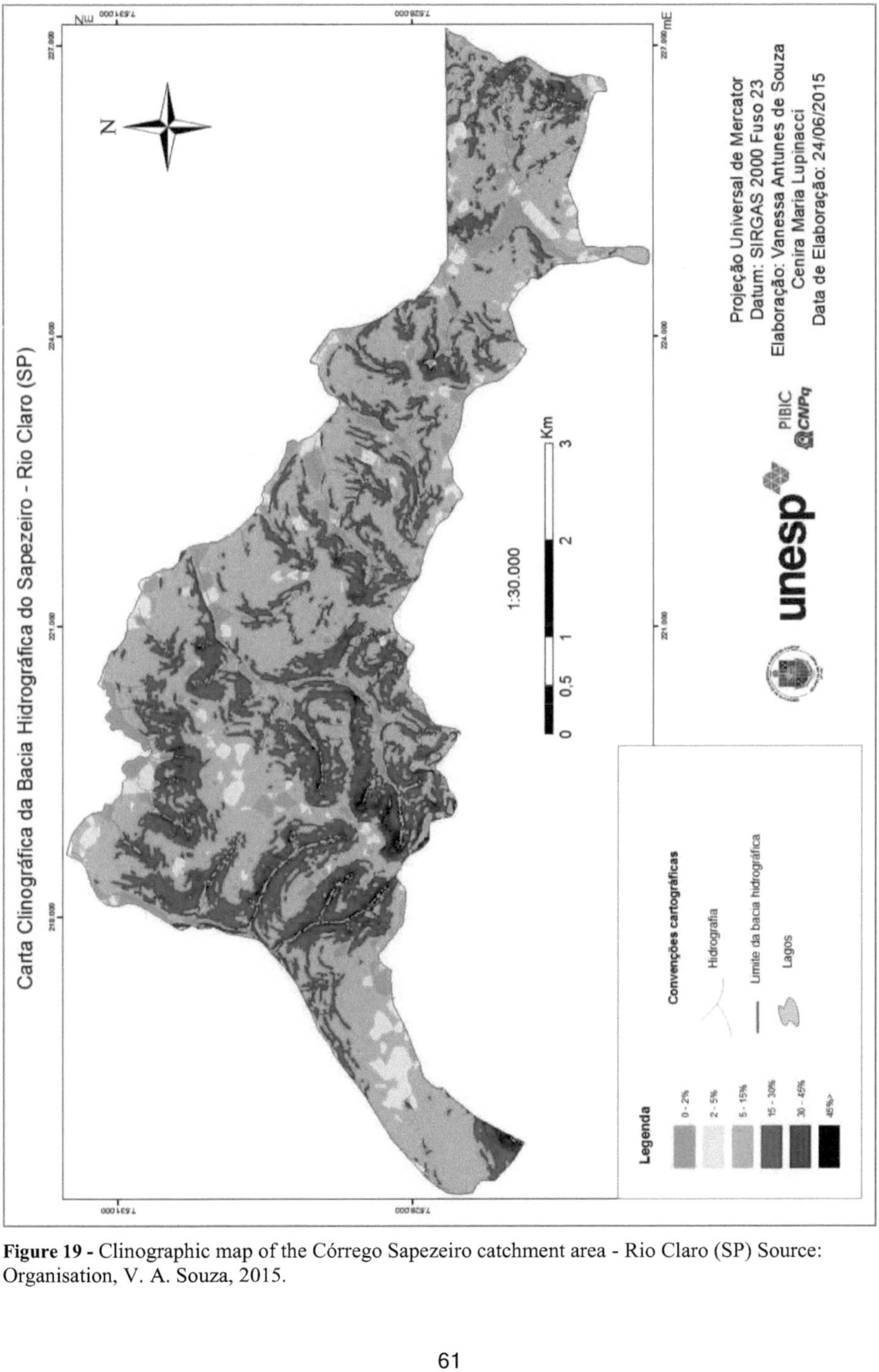

Figure 19 - Clinographic map of the Córrego Sapezeiro catchment area - Rio Claro (SP) Source: Organisation, V. A. Souza, 2015.

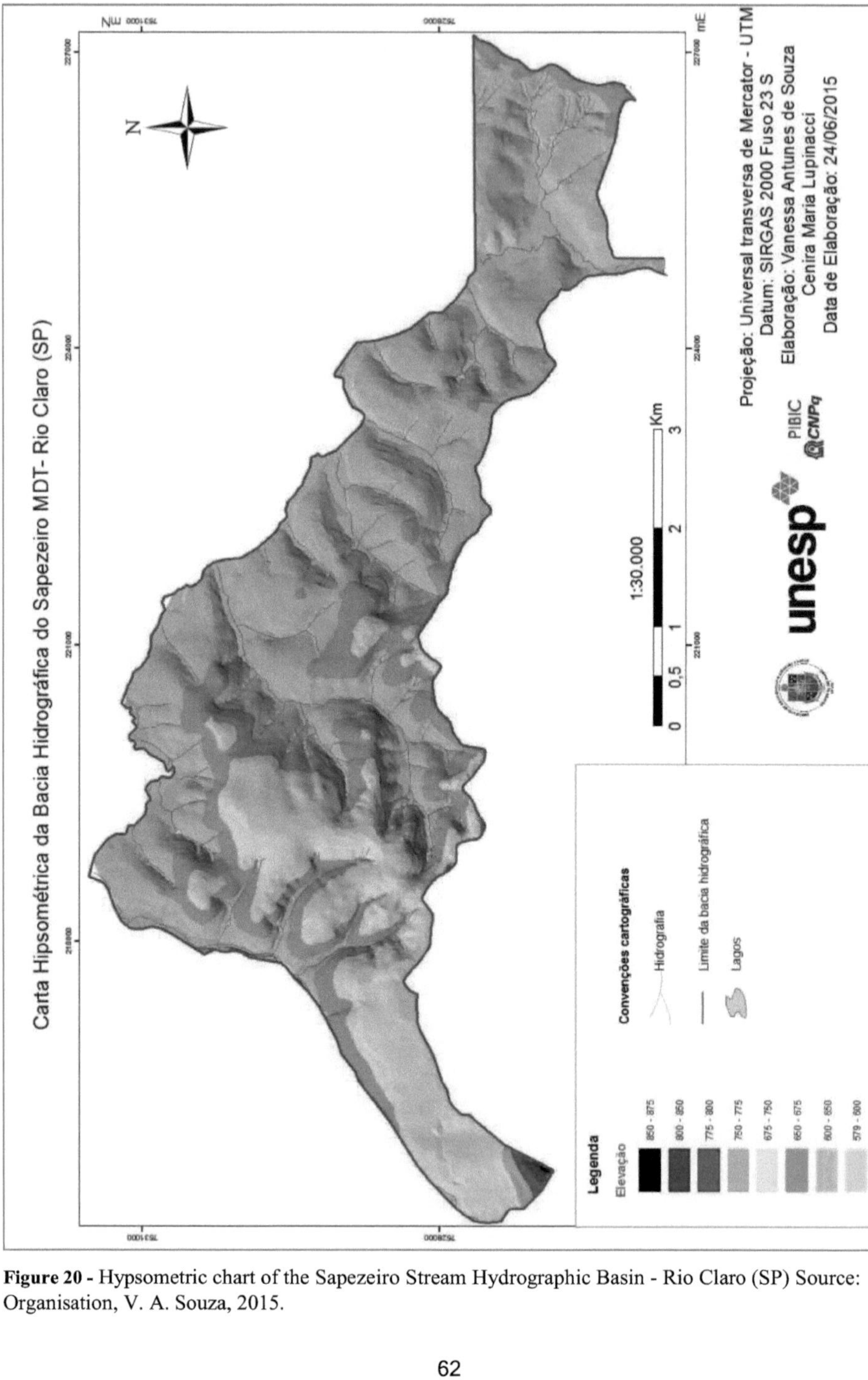

Figure 20 - Hypsometric chart of the Sapezeiro Stream Hydrographic Basin - Rio Claro (SP) Source: Organisation, V. A. Souza, 2015.

Firstly, we will analyse the northern sector of the river basin, where sugarcane monoculture has predominated, interrupted by small patches of forest in the areas of springs and on the banks of river channels (riparian forests). The properties in this area are considered medium-sized[19] . There is the use of machinery and technical assistance on site.

In this sector, the topography is undulating according to the EMBRAPA (1995) classification, marked by altitudes ranging from 600 to 750 metres, with slopes concentrated between 5-15% and 15-30%, with the steepest areas preserved with woodland around the drainage. In this sector, there is a predominance of dystrophic red-yellow Argissols with an abrupt character. The Pirambóia Formation is the predominant lithology. This gives the area very sandy, friable soils that are susceptible to erosion (Figure 21).

Figure 21 - North Sector/Evidence of the slope of the area and the sandy soils on the road. (a) the presence of sand deposits in its surroundings. (b) area of intense sugar cane activity. Source: Fieldwork, 2015.

In the Centre-North sector, there is a predominance of pastures (clean and dirty), forestry and patches of riparian forest, and, to a lesser extent, annual crops, wetlands, fish farming and poultry farming. Most of the properties are over 50 hectares in size and are considered medium to large. There are a few small properties, managed by the owner and his family, with subsistence farming, the use of chemical fertilisers, and land rented out for livestock farming; some are used by the owners as leisure areas at weekends, and they do not live there.

---

[19]Fiscal Module of the Municipality of Rio Claro - SP/ Fiscal Module =14 ha (CATI, 2008).
Smallholdings - rural property smaller than 1 fiscal module
Small property - rural property of 1 to 4 fiscal modules;
Medium property - rural property of 4 to 15 fiscal modules;
Large rural property with areas greater than 15 fiscal modules.
(Law 8.629, of 25 February 1993, INCRA, 2016).

The topography in this sector is undulating, with altitudes ranging from 675-750 metres, with a slope of between 5-15% and 15-30%, and the contact between the slopes and the valley bottoms is marked by a gentle topographic break.

In this area there are several morphological features that show a lot of erosive activity at the site, which are strictly related to anthropogenic activities. Thus, there are topographic breaks in the slopes, erosion furrows, gullies, exposed soil and poorly maintained terraces (Figure 22). There are erosion furrows on the slopes due to cattle trampling and gullies in the areas where rainwater flows disperse. At the bottom of the valley, the river channel (stream) was dammed and a weir formed.

Figure 22 - North-Central Sector/(1) Degraded, sloping pasture area, (a) topographic breaks in the slopes, (b) contact with preserved tree vegetation (riparian forest); (2) erosion furrows and gullies in the slopes due to cattle trampling; (3) gullies in the areas where rainwater flows disperse; (4) damming of the river channel (stream) with the formation of a weir (c) forested areas (d).
Source: Fieldwork, 2015.

In this area, there is a predominance of dystrophic Red-Yellow Argisols with an abrupt character, with Quartzarenic Neosols (Quartz Sands) appearing in the higher altitude areas (675-750 metres); these are very low productivity soils, due to their low water and nutrient retention capacity, low cohesion and adhesion, and also because they are sandy and acidic soils, they are very susceptible to erosion, as evidenced in the study area. There are undifferentiated Tertiary coverings (sands and gravels) in the outflow areas (top areas) and also bands of the Pirambóia Formation.

The southern sector of the catchment area is predominantly pastureland, with small patches of riparian forest and wetlands in the flatter areas near the river channels and isolated sugar cane fields used to feed cattle. The properties are medium to small and are predominantly run by caretakers and employees who live there with their families. Of the 5 properties where information was collected in this sector, only 1 uses agricultural inputs (tractor, chemical and organic fertiliser). There is technical assistance on site, but it is geared towards livestock farming.

The topography is undulating and strongly undulating according to the EMBRAPA classification (1995), marked by altitudes ranging from 600 to 750 metres, with slopes between 15-30% and 30-45%. The Pirambóia Formation is predominant and the soil is made up of dystrophic red-yellow Argisols with an abrupt character. It has erosion furrows, gullies and wide, high terraces. Erosion containment techniques are present in the pasture areas of this sector. However, as Figure 23 and 24 show, these are not sufficient.

Figure 23 - South Sector/(1) erosion furrows and gullies (2) wide, high terraces Source: Fieldwork, 2015.

Figure 24 -South Sector (Terraces)/ (1) Visualisation of the south sector from aerial photography of the

wide, high terraces, measures designed to contain erosion in pasture areas.
Source: EMPLASA, 2011.

With regard to the Centre-South sector, it was found that this is the area with the most intense occupation, with an urban or isolated urban agglomeration according to the SEPLADEMA (2015) classification, where relations between rural landowners take place more actively. There are properties ranging from homes to smallholdings (chácaras) and small properties (4 fiscal modules).

It can be seen that the owner and his family predominate, with 36% of the 22 properties visited in this sector; 27% are caretakers, responsible for the upkeep of the property, which the owners use for leisure; in 18% of these nobody currently lives on the property, using it only as a leisure area at weekends; 9% employees, who carry out an economic activity on the property (rural labourer), whose profit goes to the owner, being salaried; 5% family members (uncle, cousin, brother-in-law and others); Tenant (lessee) of a rural residence, who doesn't produce on the property and carries out an economic activity in the city (Graph 8).

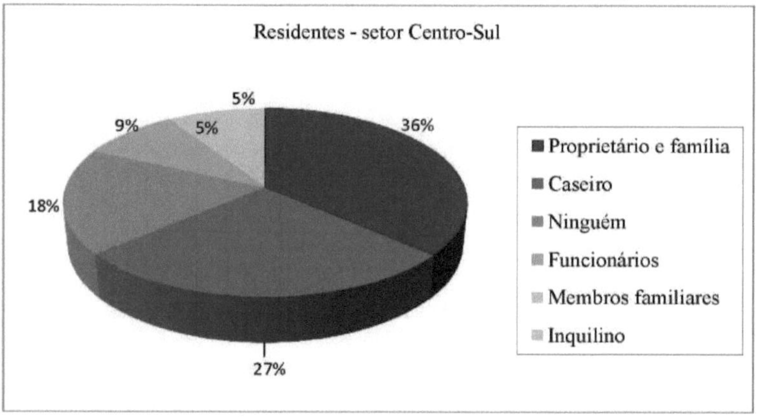

Graph 8 - Residents in the Centre-South sector of the Córrego do Sapezeiro catchment area Source: Personal archive, 2015.

The Centre-South sector is home to churches (Catholic and Evangelical), the neighbourhood football club and the health centre (UPA[20]), the latter of which was set up in the old FEPASA railway station, the same place where administrative meetings between the neighbourhood's population and Rio Claro City Hall officials eventually take place (Figure 25).

[20] Emergency Care Unit.

Figure 25 - - Urban agglomeration of the Centre-South sector/(1) Visualisation of the urban agglomeration of the Centre-South sector from aerial photography
Source: EMPLASA, 2011.

With regard to land use and occupation, the Centre-South sector has pastures (clean) and dirty pastures, which are poorly maintained along the rivers, with little riparian forest and small isolated patches. There are also erosion-depositional processes in the watercourse, the formation of sand banks in the channel and erosion furrows on the banks (Figure 26).

Figure 26 - (1) Cabeça River/(a) forest; (b) citrus grove; (c) dirty pastures; (d) erosion on the bank of the (1) Cabeça River.
Source: Fieldwork, 2015.

Pastures are interrupted by annual crops, citrus farming, fish farming, sugar cane and wetlands near rivers. Agricultural inputs such as corrective, chemical and organic fertilisers are used, as well as machinery such as tractors (37% of the owners in this sector have them), seed drills, tampers and chippers. Of the properties visited in this sector, 23% have technical assistance and 18% have financing for agricultural activities.

Some of the owners maintain a small artisanal production of cheeses and sweets, which are sold on site, as well as handicrafts such as napkins, jewellery, scarves and others, produced by women who take part in the Senac Social Network programme[21] in Itirapina - SP, a neighbouring municipality.

As far as the physical aspects of the sector are concerned, the topography is gentle to moderately undulating, according to the EMBRAPA classification (1995), marked by altitudes ranging from 600-675 metres, with slopes between 0-2% and 5-15%.

The Pirambóia Formation occurs predominantly in this sector, and the soils that make up this sector are dystrophic red-yellow Argisols with an apricot character and red-yellow Latossolos, alic, moderate A with a medium texture. The latter requires constant correction and fertilisation for good productivity due to its concentration of aluminium, has a high sand content and its medium texture gives it good permeability and low water retention.

Thus, such soils dry out quickly after rain or irrigation and are also more prone to water stress in summer and more susceptible to erosion.

The western sector of the catchment area is an area of contrasting landscapes, where the morphological aspects of the hilly terrain, with broad hills with gently convex slopes, gently sloping plateaus, with a predominance of sugar cane monoculture and pastures, contrast with the relief of the sandstone-basalt cuestas and the patches of forest and forestry. The properties are considered medium-sized and the residents are predominantly owners and their families.

The forest areas only occur as remnants in Permanent Preservation Areas, places that are difficult to access (mainly on the slopes of the cuestas), where the use of machinery is prevented. These areas may not be productive in terms of agricultural production, due to their limitations and impediments, but they can perform other activities aimed at environmental education and tourism (Figure 27).

21 National Commercial Apprenticeship Service - Regional Administration of the State of São Paulo.

Figure 27 - West Sector/(1; 2) West Sector (a) Forestry and (b) woodland on the slopes of the **Cuestas.**
Source: Fieldwork, 2015.

The topography is moderately undulating, where sugar cane monoculture and forestry activities occur, and steep near the front of the sandstone-basalt cuestas, according to the EMBRAPA classification (1995).

The altitudes of this sector vary from 675-875 metres, with slopes ranging from 5-15% to over 45%. The Pirambóia Formation and corridors of undifferentiated cover (sands and gravels) occur.

The western sector has the most varied types of soils, with the presence of Red-Yellow Latosols, alkalic, moderate A, medium texture; patches of Quartz Sands (Quartzarenic Neosols), deep, alkalic, medium texture; Terra Roxa (Purple Latosol), eutrophic (high fertility) and dystrophic (low fertility), moderate A, clayey texture; soils from the decomposition of diabase; and the Litholic Neosols, new soils, eutrophic and dystrophic, moderate A, indiscriminate texture, sandstone substrate Botucatu-Pirambóia and clayey texture, basalt and diabase substrate.

Finally, in the eastern sector of the catchment area, there are medium-sized properties managed by owners and employees, especially in the pottery sector. In this sector, many properties are leased for sugar cane, which is the predominant use in this area (Figure 28), interrupted by small patches of forest in the areas of springs and on the banks of river channels (riparian forests), poorly maintained pastures (dirty and clean), wetlands near the Cabeça River and sand and clay mining activities (Figure 29). The properties use agricultural inputs such as tractors, correctives and chemical fertilisers for sugarcane plantations.

Figure 28 - East Sector/(a) Sugarcane monoculture, (b) pastures and dam and (c) woodlands Source: Fieldwork, 2015.

Figure 29 - East Sector (Mining area)/(1) East Sector, visualisation of erosion features, exposed soil and topographic break and (2) mining and pottery areas.
Source: Fieldwork, 2015.

The topography in this sector is flat according to the EMBRAPA classification (1995), marked by altitudes ranging from 600-675 metres, with slopes concentrated between 0-2% and 5-15%. There is a predominance of dystrophic red-yellow Argisols with an abrupt character and a small patch of red-yellow Latosol, atic, moderate A, medium texture.

The Corumbataí Formation (claystones and siltstones) occurs predominantly in this sector, being interrupted by corridors of Recent Quaternary Deposits (sands, clay and gravel) around the

Cabeça river channel. There is also a small patch at the eastern end of the catchment area of the Rio Claro Formation (claystones, conglomerates and sandstones) and the Pirambóia Formation.

The Corumbataí Formation provides raw materials for the region's potteries. The Pirambóia Formation supplies raw materials to the glass and foundry industries. Mining areas are found in these units, generating many erosive features, exposed soil, topographic breaks and silting up of the region's rivers (Figures 29 and 30).

Figure 30 - East Sector (Aerial photography) / (1) Visualisation from aerial photography, areas of sand mining and resulting (a) silting of the river channel and (2) areas of clay mining and erosion features. Source: EMPLASA, 2011.

Below we will present the final considerations of the analysis of the physical attributes of the Córrego do Sapezeiro catchment area, based on the data obtained in this research and considering the potential multifunctionality of the landscape contacted.

CHAPTER 4

FINAL CONSIDERATIONS

The analysis carried out in the Córrego Sapezeiro watershed made it possible to characterise the multifunctional potential of the small rural properties in the Itapé neighbourhood located in the respective study area, to understand the different forms of land use and occupation that have been taking place in the area, their limitations and difficulties in terms of both physical and socio-economic aspects, which imply local development, especially for small family owners.

With regard to socio-economic data, it was found that the local residents are predominantly landowners and their families, who carry out subsistence farming and other economic activities in the city; another very important figure was that of the caretakers, who are responsible for maintaining the properties, where many landowners use the place as a leisure environment at weekends and on holidays, carrying out activities in the city, such as teaching, micro-entrepreneurship, bricklaying, among others. Many of these properties are rented out for sugar cane and livestock farming.

The properties were bought and abandoned, in many cases due to the precariousness of the area and because they didn't consider it profitable to produce, either due to the difficulty of using machinery in steep areas or the sandy soils that are very susceptible to erosion (a characteristic of the area), lack of technical assistance, difficulty in getting finance to buy machinery, fertilisers, seedlings and other agricultural inputs.

This is coupled with the precarious infrastructure of the neighbourhood, the lack of basic sanitation and essential services, as well as the isolation of the region and the lack of attractiveness and opportunities to keep the population in the area.

The area studied is supplied by the Córrego Sapezeiro basin and the Cabeça river, which are rivers of great local importance, as they are supplied by mines, springs and wells. The Pirambóia Formation appears predominantly in the area, with sandy (red) layers, medium to fine-grained, coarse and conglomeratic sandstones and clay fraction, high permeability. The Corumbataí formation has grey, purplish or reddish claystones, shales and siltstones, carbonates, siltstones and fine sandstones. Recent deposits in small isolated patches, sands and gravels and colluvial deposits (IPT, 1981).

The soils are predominantly A moderate, sandy/medium dystrophic Argissolo Vermelho-Amarelo (PVA) and A moderate, medium textured, dystrophic Latossolos Vermelho-Amarela. There are small patches of deep, alkaline, moderate A quartz sand (AQ). The study area has a predominantly 15% slope, which, according to the definition of relief proposed for slope classes by Ramalho Filho & Beek (1995, apud EMBRAPA, 2004), is class D, whose relief is considered to be undulating.

Analysing the physical attributes of the landscape together with the type of use and occupation that has been taking place in the watershed has made it possible to understand how much the limitations imposed by physical conditions interfere with local agricultural productivity, especially in relation to small rural landowners.

In the basin there is a predominance of pastures (clean and dirty pastures) in land use; this activity is carried out on poor, sandy soils that are highly susceptible to erosion. In most of the study area, this activity has been carried out without proper planning, disregarding the limitations and capacity of the land, thus severely degrading it.

As a result, there are many erosive features caused by grazing, such as furrows and gullies, as well as the disorderly occupation of this activity in areas of riparian forest, steep slopes and springs, strongly compromising the local physical environment and jeopardising the maintenance of water resources. There are many dirty pastures in the region. As a result, there is low productivity, which jeopardises socio-economic development.

With regard to woodlands, we can see important areas in the different sectors of the basin, comprising 23% of the total area, but these are fragmented even on steep slopes and on the banks of river channels (riparian woodlands). As a result, they do not form ecological corridors along the different sectors of the river basin, harming the flora and fauna.

Sugarcane is a significant crop in the study area, occupying extensive areas of up to 15 per cent slope, sandy and impoverished soils. In the areas occupied by this activity, many evolving erosion features were recorded, especially erosion furrows and gullies.

Mining is another activity that takes place in the region, and although it has a small share in land use and occupation, it causes major environmental impacts. This activity occurs predominantly in the eastern part of the catchment area, giving the local landscape many erosive features, exposed soil, topographical breaks and even silting up the rivers in this sector.

Another aspect highlighted in the field and in photointerpretation was the removal of vegetation and agricultural activities by gas pipelines that extend from the north to the southeast of the catchment area, with strips of exposed soil, erosion furrows, dam embankments and the removal of vegetation along the 4.8 kilometre length of the pipelines.

In addition, other activities such as forestry, fish farming, poultry farming (farms) and annual crops (manioc, maize, beans, mangoes, avocados and others) were recorded with a smaller share of the total area of land use and occupation, but which on the other hand play a significant role in the subsistence of local families, especially in the Centre-South sector, where they occur more expressively, and this diversity is one of the multifunctional potentials of the region to be considered for local development, considering that many already develop by-products such as cheese and artisanal jam. In addition to the handicrafts produced by the women who take part in the Senac Social Network programme in Itirapina - SP, a neighbouring municipality.

So, thinking about alternatives to minimise these problems; considering that the area is currently in the Rio Claro Macrozone of Preservation and Sustainable Use, conservation practices can offer the opportunity to cultivate the soil without significantly impoverishing it. With proper soil management, combined with land use that respects its limitations and considers its potential, we can be much more productive and for a longer period of time. To this end, the implementation of payments for environmental services (a proposal established in the New Master Plan, 2015) has made it possible to recover degraded areas and form ecological corridors, preserving water sources, flora and fauna.

Taking into account the infrastructural, social and physical limitations found in the Itapé rural neighbourhood, we must consider the possibility of developing "new functions" for the development of local families.

Therefore, expanding the possibilities for the permanence and development of the neighbourhood could be dealt with by taking advantage of the potential of the Itapé neighbourhood landscape, which would provide opportunities for income generation, without losing sight of the sustainable use of natural resources, the maintenance of biodiversity, the preservation of ecosystems, historical, cultural and environmental heritage. Thus, as Complementary Law No. XX/2015 Section II, Art. 29 establishes, the following are the main objectives of the Environmental Preservation Areas:

I-Conserve, recover and restore the environment and natural habitat of the municipality;

II-Reserves the integrity of fragments of native biomes that still exist in the municipality;

III-Control and supervise the uses installed in the vicinity of the Cabeça and Passa Cinco rivers, respecting those permitted in the Environmental Protection Areas (APAs);

IV-Restrict urban uses and the process of anthropisation of the area;

V-Establish possibilities for low-impact rural and tourist uses;

VI-Prioritise payment programmes for environmental services;

VII-Protecting and restoring water sources.

(Masterplan, Supplementary Law 201/2015, art. 29).

Reflecting on initiatives that aim to enhance, revitalise and conserve the landscape of the study area, we can think of the rescue of historical, cultural and environmental heritage, both in terms of material and immaterial aspects, considering that:

> Historical heritage can be understood as the visible factors in the landscape that have been lost over time, such as architecture, documents, utensils, among others, which are important for understanding its historical identity. Cultural heritage is present in the invisible facts of the landscape, which tell the story through customs, gastronomy, religion, legends, festivals, among others. The environmental heritage that involves the interrelationship between man and natural resources, which in many cases is the domain of man over nature, causing conflicts and preservationist decision-making (PPGPPC, 2011 apud DAMBRÓS, C.;FERREIRA, D. A. O., 2012, p.4).

The revitalisation of the area's historical and architectural heritage and cultural traditions with the creation of a Centre for the Appreciation of Cultural and Historical Heritage in the old FEPASA railway station building (relocation of the UBS to another, more suitable building on the site), thus creating a space that can bring together documents, utensils, that can recover history, as well as recovering customs, music, stories, festivals, gastronomy, handicrafts, community awareness actions, encouraging culture.

Encouraging events and entertainment activities promoted by the chapel of the Divine Saviour together with the community, such as the feast of the patron saint of the Saviour, Saint Anthony, and others mentioned by some interviewees, such as Christmas dances and festivities.

The formation of family farmers' cooperatives, with the aim of generating opportunities and income for the members, through the processing and improvement of agricultural products, the creation of craft groups and the participation and creation of fairs both in the town, which is already being developed and organised by the town hall, and the possibility of developing them locally, would also be welcome. It would also increase the participation of residents in drawing up measures aimed at local economic and social development.

The creation of environmental education both for the community and activities aimed at receiving visitors, an itinerary of activities aimed at both ecotourism and agrotourism, with the participation of the community, especially children and young people.

Therefore, the co-operation and participation of the public administration with the rural population, encouraging infrastructure improvements and developing policies aimed at expanding the possibilities and permanence of the rural population is fundamental to achieving some success.

Therefore, it should be considered that certain areas such as Itapé may lose their competitive capacity at the production level, however, they develop other functions such as: recreational support, nature conservation, environmental quality, aesthetics and preservation of cultural and architectural heritage, since not all rural spaces will be productive spaces.

CHAPTER 5

REFERENCES

AUED, B. W.; PAULILO, M. I. S. (Org.) A agricultura Familiar. Florianópolis: Insular, 2004.

ABRAMOVAY, R. Subsidies and Multifunctionality in European Agricultural Policy. Revista de Economia e Sociologia Rural. São Paulo, vol. 40, n°2, p. 235-264.

BERTONI, J. Conservação do Solo, São Paulo: Ícone, 2012.

CARNEIRO, M. J.; MALUF, R. S. (org.) Beyond production: multifunctionality and family farming. Rio de Janeiro/Brasília: Ed. Mauad/Nead, 2003.

CARVALHO, M. L. da S.; LUCAS VENTURAS, M. R. D; HENRIQUES DAMIÃO, P. Multifunctionality in Agriculture and its Valences. Available at: < http://www.sper.pt/IICER/pdfs/Tema1/L_Carvalho.pdf> Accessed on: 06/09/2012

CAETANO-CHANG M. R.; WU F.T. Diagenesis of Sandstones of the Pirambóia Formation in Central-Eastern São Paulo, São Paulo, UNESP, Geociência, v. 22, N. Especial, p. 33-39, 2003.

CAZELLA, A. A.; BONNAL, P.; MALUF, R. S. (org.) Agricultura familiar: multifuncionalidade e desenvolvimento territorial no Brasil. Rio de Janeiro: Mauad X, 2009.

CEPAGRi Available: http://www.cpa.unicamp.br/outras-informacoes/clima- dos-municipios-paulistas.html Accessed on: 22/07/2015.

DAMBRÓS, C; FERREIRA, D. A. O. Multifunctionality of the Landscape: the role in the management of historical, cultural and environmental heritage in family farming areas In: XXI ENCONTRO NACIONAL DE GEOGRAFIA AGRÁRIA, 2012, Uberlândia-MG, UFU, 2012,

DE BIASI, M. A Carta Clinográfica: os métodos de representação e sua confecção Revista do Departamento de Geografia, São Paulo, n. 6, p. 45 - 60, 1992.

FERNANDES, V. O. Implicações da Adoção do Referencial Geodésico SIRGAS 2000 na Cartografia em Escala Grande. f. 141. Tese (Doutorado) - Universidade Federal de Santa Catarina.

FERREIRA, D. A. O. Mundo Rural e Geografia. Agrarian Geography in Brazil 1930-1990, São Paulo: Editora UNESP, 2002.

FROEHLICH, J. M.; DULLIUS, P. R.; PIETRZACKA, R. A Multifuncionalidade do Espaço Rural na Região Central do Rio Grande Do Sul - Dados Gerais. Paper presented at the XLIII Congress of the Brazilian Society of Rural Economics and Sociology 2005. Available at: < http://coral.ufsm.br/desenvolvimentorural/textos/sober%202005%203.pdf> Accessed on:

11/10/2012

GALVÃO, M. J.; VARETA, N. D. The multifunctionality of rural landscapes: a a tool for development. Cadernos Doctoral Programme in Geography. Faculty of Letters of the University of Porto - FLUP, Portugal 2010.
Available at:< http://ler.letras.up.pt/uploads/ficheiros/8282.pdf.> Accessed on: 06/09/2012

GRAZIANO DA SILVA, J. O novo rural brasileiro. New Economy. Belo Horizonte, v. 7, n.1; pp43-81, 1998.

IPT. Technological Research Institute of the State of São Paulo. Geological Map of the State of São Paulo. São Paulo: 1981b., v. 2, Serie Monigrafias, Scale 1:50.000.

GEOLOGICAL INSTITUTE (SECRETARIAT OF AGRICULTURE AND SUPPLY) OF THE STATE OF SAO PAULO. Geological Sheet of Rio Claro. Scale 1:50,000, 1986.

KAGEYAMA, A. Rural e Ruralidades in Desenvolvimento Rural Conceitos e Explicações ao Caso Brasileiro. Porto Alegre: Editora da UFRGS de Pós Graduação em Desenvolvimento Rural, 2008.

LEPSCH, I.F. Formação e Conservação dos Solos, São Paulo: Oficina de Textos, 2002.

SEMI-DETALISED SURVEY OF THE STATE OF SÃO PAULO - São Carlos squares, scale 1:100,000, 1991.

LEPSCH, LF." Manual for the Survey of the Physical Environment and Land Classification in the Use Capacity System". Campinas: SBCS, 1983.

LOMBARDI NETO; PEREIRA, Avaliação da aptidão agrícola das terras: proposta metodológica, Jaguariúna: Embrapa Meio Ambiente, 2004.

MACHADO, A. G.; FIGUEIREDO, R. S.; CAUME, D. J.; OLIVEIRA NETO, O. J. Multifunctionality and Pluriactivity as Development Alternatives for Family Farming in Brazil. Desafio: Revista de Economia e Administração, v. 9, n. 17, p. 19-30, 2008.

MELO, V. L. M. O. The Landscape from the Perspective of New Geographical Approaches. Available at : <http://mazinger.sisib.uchile.cl/repositorio/ap/arquitectura y urbanismo/> Accessed on:05/11/2012

MESQUITA, P. E. Multifunctionality and Non-Trade Concerns: Implications for Agricultural Negotiations at the WTO. Brasília: Funag, 2005.

MULLER, J. M. Multifunctionality of Agriculture and Family Farming: The Reconstruction of Rural Spaces in Perspective. Florianópolis: INCRA, 2005
Available at :
http://www.cnpat.embrapa.br/sbsp/anais/Trab Format PDF/73.pdf

Accessed on: 16/01/2013

N. GUIOMAR; J. P. FERNANDES; N. NEVES Modelo de Analise Espacial para Avaliação do Carácter Multifuncional do Espaço. Proceedings of the III Congress of Rural Studies (III CER), Faro, University of Algarve, 2007.
Available
at: <http: //www. sper. pt/IIICER/Comunicacoes/NGuiomar et al .2 com.pdf> Accessed on: 04/10/2012

OLIVEIRA, L.; MACHADO, LUCY M. C. P. Landscape Perception as a Geographical Research Methodology. Available at
http://observatoriogeograflcoamericalatina.org.mx/egal2/Teoriaymetodo/Metodo logicos/10.pdf
Accessed: 04/10/2012

PINTO-CORREIA, T. Multifunctionality of the Rural Landscape: New Challenges for its Analysis. Inforgeo. July, ed. Portuguese Association of Geographers. 2007. Available at :
http://www.apgeo.pt/files/docs/Inforgeo 20&21/Inforgeo%2020&21%20Pagina s 067 071.pdf
Accessed: 09/10/2012

PINTO-CORREIA, T.; BARROSO, F.; MENEZES, H.. TAVEIRA, R. Que Multifuncionalidade na Paisagem Rural: Sinergias e Conflitos entre a

Agriculture and Non-Productive Functions. Application to the Municipality of Castelo de Vide , Faro, University of Algarve, 2008. Available at:
http://dynamo08.files.wordpress.com/2008/07/sper-iii-nov07-291007-final.pdf Acesso em: 09/10/2012

SOARES, A. C. Multifunctionality of Family Farming in International Trade, Food Security and Family Farming. ActionAid Brasil, 2001. Available at :
http://www.fase.org.br/projetos/clientes/noar/noar/UserFiles/20/File/Outras%20 publicacoes/a%20multifuncionalidade%20da%20agricultura%20familiar.pdf Accessed on: 28/9/2012

SCHIER, R. A. Trajectories of the Concept of Landscape in Geography. Curitiba: Raega, n. 7, 2003.

More Books!

yes

I want morebooks!

Buy your books fast and straightforward online - at one of world's fastest growing online book stores! Environmentally sound due to Print-on-Demand technologies.

Buy your books online at
www.morebooks.shop

Kaufen Sie Ihre Bücher schnell und unkompliziert online – auf einer der am schnellsten wachsenden Buchhandelsplattformen weltweit! Dank Print-On-Demand umwelt- und ressourcenschonend produzi ert.

Bücher schneller online kaufen
www.morebooks.shop

info@omniscriptum.com
www.omniscriptum.com

OMNIScriptum

MIX
Papier aus verantwortungsvollen Quellen
Paper from responsible sources
FSC® C105338

Printed by Books on Demand GmbH, Norderstedt / Germany